ELEMENTARY ELECTRIC POWER AND MACHINES

ELLIS HORWOOD SERIES IN ELECTRICAL AND ELECTRONIC ENGINEERING

Series Editors: PETER BRANDON, Professor of Electrical and Electronic Engineering, University of Cambridge

P. R. ADBY, Department of Electronic and Electrical Engineering, University of London, King's College

HANDBOOK OF ELECTRONIC CIRCUITS
G. J. SCOLES, formerly English Electric Valve Company, Chelmsford, Essex

APPLIED CIRCUIT THEORY: Matrix and Computer Methods
P. R. ADBY, Department of Electronic and Electrical Engineering, University of London, Kings's College

DIFFRACTION THEORY AND ANTENNAS
R. H. CLARKE, Lecturer, Department of Electrical Engineering and
J. BROWNE, Professor of Electrical Engineering, Imperial College of Science and Technology, University of London

HANDBOOK OF RECTIFIER CIRCUITS
G. J. SCOLES, formerly English Electric Valve Company, Chelmsford, Essex

DIGITAL AND MICROPROCESSOR ENGINEERING
S. J. CAHILL, Senior Lecturer at the School of Electrical and Electronic Engineering, Ulster Polytechnic

POST-DIGITAL ELECTRONICS
F. R. PETTIT, Head of the Computing Teaching Centre, Oxford

BOND GRAPHS FOR MODELLING ENGINEERING SYSTEMS
ALAN BLUNDELL, Senior Lecturer, Department of Systems and Control, Coventry (Lanchester) Polytechnic

NOISE IN ELECTRONIC DEVICES AND SYSTEMS
M. J. BUCKINGHAM, Royal Aircraft Establishment

ELEMENTARY ELECTRIC POWER AND MACHINES
P. G. McLAREN, University Engineering Department, University of Cambridge

QUANTITATIVE ELECTRON-PROBE MICROANALYSIS
V. D. SCOTT and G. LOVE, School of Materials Science, University of Bath

PRINCIPLES OF COMPUTER COMMUNICATION NETWORK DESIGN
J. SEIDLER, Institute of Information Science, Technical University of Gdansk, Poland

NON-LINEAR AND PARAMETRIC CIRCUITS: Principles, Theory and Applications
F. KOURIL and K. VRBA, Technical University of Brno, Czechoslovakia

ELEMENTARY ELECTRIC POWER AND MACHINES

P. G. McLAREN, B.Sc.(1st Class Hons.), Ph.D., M.I.E.E., M.I.E.E.E.
Lecturer in Engineering, University of Cambridge
and Fellow of Churchill College

ELLIS HORWOOD LIMITED
Publishers · Chichester

Halsted Press: a division of
JOHN WILEY & SONS
New York · Chichester · Brisbane · Toronto

First published in 1984 by
ELLIS HORWOOD LIMITED
Market Cross House, Cooper Street, Chichester, West Sussex, PO19 1EB, England

The publisher's colophon is reproduced from James Gillison's drawing of the ancient Market Cross, Chichester.

Distributors:

Australia, New Zealand, South-east Asia:
Jacaranda-Wiley Ltd., Jacaranda Press,
JOHN WILEY & SONS INC.,
G.P.O. Box 859, Brisbane, Queensland 40001, Australia

Canada:
JOHN WILEY & SONS CANADA LIMITED
22 Worcester Road, Rexdale, Ontario, Canada.

Europe, Africa:
JOHN WILEY & SONS LIMITED
Baffins Lane, Chichester, West Sussex, England.

North and South America and the rest of the world:
Halsted Press: a division of
JOHN WILEY & SONS
605 Third Avenue, New York, N.Y. 10016, U.S.A.

British Library Cataloguing in Publication Data
McLaren, P.G.
Elementary electric power and machines. –
(Ellis Horwood series in electrical and electronic engineering)
1. Electric apparatus and appliances
I. Title
621.31'042 TK452

Library of Congress Card No. 83-26666

ISBN 0-85312-269-5 (Ellis Horwood Limited – Library Edn.)
ISBN 0-85312-715-8 (Ellis Horwood Limited – Student Edn.)
ISBN 0-470-200-54-5 (Halsted Press – Library Edn.)
ISBN 0-470-200-57-X (Halsted Press – Student Edn.)

Printed in Great Britain by R. J. Acford, Chichester.

Table of Contents

Foreword

All engineers require some knowledge of electrical power and machines. The majority will require this knowledge as users or future users of such facilities with relatively few becoming involved in the design of electrical power devices or circuits. This text is expressly aimed at the non-specialist and as a result contains little in the way of design detail. Adequate references are given, however, for the reader who wishes to enquire beyond the scope of this particular book.

After teaching this topic in several British universities it is the author's experience that good engineers wish to understand the basic mode of operation of any device they need to use. This is particularly true at student level. Traditional textbooks on the subject of electrical machines have made this ideal difficult to attain since such books have been written with the specialist rather than the non-specialist in mind. Nowhere is this division between specialist and non-speciliast more evident than in the sections of such textbooks dealing with machine windings. Yet no respectable text can afford to ignore this topic any more than a book on transistor circuits could afford to ignore the physics of the p–n junction.

Chapters 1 and 2 cover the basic circuit theory and field theory necessary to explain the steady state behaviour of single and three-phase power devices. Chapter 3 deals with the power transformer and autotransformer while Chapter 4 introduces the reader to very simple machine windings. Hopefully the reader will emerge from this with sufficient enthusiasm to tackle Chapters 5, 6 and 7 which cover the steady state operating characteristics of the synchronous machine, the induction motor and the d.c. machine respectively. The emphasis throughout Chapters 3, 5, 6 and 7 is on understanding the derivation of the equivalent circuit for each machine type. Many worked examples are included to aid the exposition and although transient analysis is not covered starting procedures for each machine are briefly explained. The most complex analysis is that concerned with the single-phase induction motor in Chapter 6. This is such a common machine that it must be given adequate coverage despite its complexity.

The final chapters, 8 and 9, are of a descriptive nature covering the less common and novel machine types and the rapidly expanding topic of power electronics. The combination of machines and power electronics into 'drive systems' with the ubiquitous microprocessor doing its several bits in the control process, is emerging as an important subject area for any engineer going into manufacturing industry. Chapter 9 does no more than open the door to this field of study but references are given to more specialist books and publications for those who wish to pursue this topic further.

In conclusion the author wishes to thank his long-suffering colleagues and family against whom he used the excuse of writing this book to escape from a multiplicity of tasks both professional and domestic. In particular he wishes to thank Dr John Hill and Professor Peter Brandon for their helpful comments on the original draft and Mrs Mavis Barber for performing the near-miracle of translating his handwriting into typescript.

Cambridge. July 1983 Peter G. McLaren

List of principal symbols

A	abbreviation for Amperes; or abbreviation for Area – see context.
a	number of parallel paths through armature winding (Chapter 7).
abc	letters used together to designate the three lines in a three-phase system. RYB also used.
AT	units of m.m.f., Ampere-Turns.
B	magnetic flux density; or susceptance — see context
C	capacitance.
D	electric flux density.
e	instantaneous e.m.f.
$\hat{E}$	amplitude of sinusoidal e.m.f. $e = \hat{E} \sin \omega t$.
$\underline{\hat{\mathbf{E}}}$	rotating vector representing $\hat{E} \sin \omega t$.
$\bar{E}$	phasor representing sinusoidal e.m.f.
$\underline{E}$	r.m.s. (root-mean-square) value of sinusoidal e.m.f.
$\bar{E}_{AB}$	phasor e.m.f. pointing towards point A from point B.
e.m.f.	electro motive force.
f	frequency.
fn	function of.
H	magnetic field intensity.
i	instantaneous current.
$\hat{I}$	amplitude of sinusoidal current $\hat{I} \sin \omega t$.
$\underline{\hat{\mathbf{I}}}$	rotating vector representing $\hat{I} \sin \omega t$.
$\bar{I}$	phasor representation of sinusoidal current.
$\underline{I}$	r.m.s. value of sinusoidal current.
$\bar{I}_{AB}$	phasor current flowing to B from A.
J	current density.
K_1, K, k_d, k, k^1	constants.

L	coefficient of self-inductance.
ℓ	length.
M	coefficient of mutual inductance.
m.m.f.	magneto motive force.
N	number of turns.
Nm	Newton-metres.
O.C.C.	Open-circuit characteristic (Chapter 7).
P	power.
p	number of poles.
R	resistance; or radius – see context.
RYB	see abc.
r	radius; or winding resistance – see context.
$\mathcal{R}$	reluctance of magnetic circuit.
Re()	Real part of quantity in brackets
s	slip (Chapter 6).
T	torque; or Tesla — see context
V	abbreviation for Volts; or r.m.s. value of phasor $\overline{V}$ – *see context.*
v	velocity; or instantaneous voltage – see context.
W	abbreviation for Watts.
Wb	abbreviation for Webers.
W_m	stored magnetic energy.
X	reactance.
x	leakage reactance.
$\overline{Y}$	admittance.
$\overline{Z}$	impedance.
$\int_A dA$	integral taken over area A.
$\int_V dV$	integral taken over volume V.
ϕ	magnetic flux; or phase angle – see context.
Φ	flux per pole (Chapter 7).
μ_r	relative permeability.
μ_o	permeability of free space ($4\pi \times 10^{-7}$).
ω	angular velocity or angular frequency.
$5 \angle 30°$	a phasor of length 5 units at an angle of 30° to the reference direction.
$\propto$	proportional to
$\doteqdot$, $\bumpeq$	approximately equal to

1

A.C. Circuits

1.1. INTRODUCTION

It will be assumed that the reader is familiar with the common circuit laws and theorems as applied to d.c. circuit analysis. This chapter will show how such laws and theorems can be applied to a.c. variables expressed in the form of phasors. Again it will be assumed that the reader is familiar with the arithmetic of complex numbers and the Argand diagram. Single phase and polyphase problems will then be solved in order to familiarise the reader with the correct application of a.c. circuit techniques.

1.2 VOLTAGE AND CURRENT RELATIONSHIPS IN R, L AND C

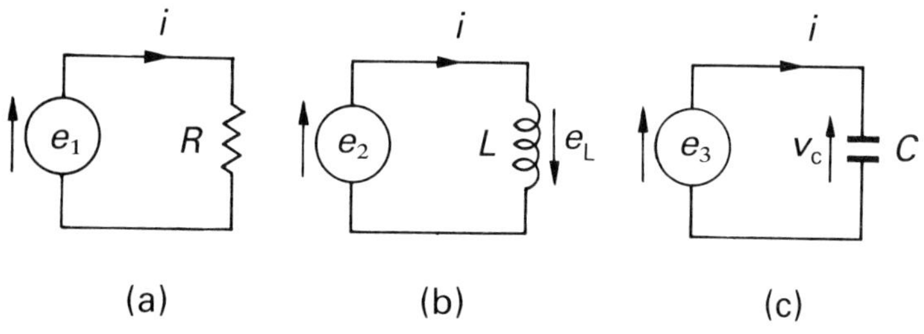

Fig. 1.1 Instantaneous current and voltage variables for (a) a pure resistance, (b) a pure inductance and (c) a pure capacitance.

Kirchhoff's second law applied to each of the circuits in Fig. 1.1 gives

$$e_1 = iR \tag{1.1}$$

$$e_2 + e_L = 0 \tag{1.2}$$

$$e_3 = v_c \tag{1.3}$$

e.m.f's have been kept on the LHS of the equations and potential differences on the RHS. Note the use of arrows on circuit diagrams rather than the + and − signs used in d.c. circuits. It remains to insert the relationships

$$e_L = -L\frac{di}{dt} \quad \text{and} \quad v_c = \frac{1}{C}\int i dt$$

to arrive at the simple equations

$$e_1 = iR \tag{1.4}$$

$$e_2 = L\frac{di}{dt} \tag{1.5}$$

$$e_3 = \frac{1}{C}\int i dt \tag{1.6}$$

If a current of waveshape shown in Fig. 1.2(a) is passed through each of the elements then the e.m.f.'s e_1, e_2, e_3 would require to be as shown in Fig. 1.2(a), (b) and (c) respectively.

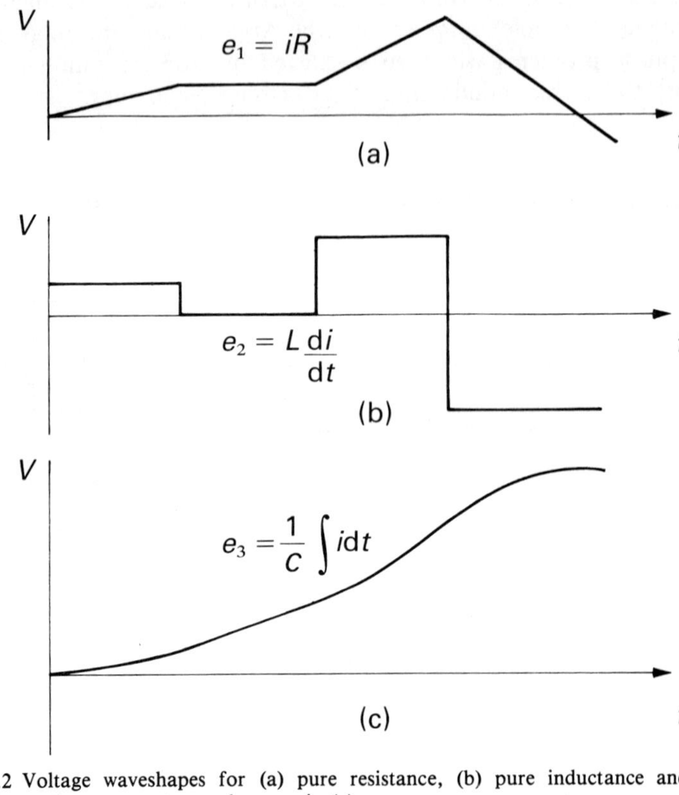

Fig. 1.2 Voltage waveshapes for (a) pure resistance, (b) pure inductance and (c) pure capacitance for a current waveshape as in (a).

In the study of a.c. power problems we are concerned with currents which are sinusoidal functions of time. Putting $i = \hat{I} \sin \omega t$ into equations (1.4), (1.5) and (1.6) gives

$$e_1 = R\hat{I} \sin \omega t \tag{1.7}$$

$$e_2 = \omega L\hat{I} \cos \omega t \tag{1.8}$$

$$e_3 = -\frac{\hat{I}}{\omega C} \cos \omega t \tag{1.9}$$

Fig. 1.3 shows the current waveform and the corresponding e.m.f's e_1, e_2 and e_3. Note that all the variables have a sinusoidal waveshape but that

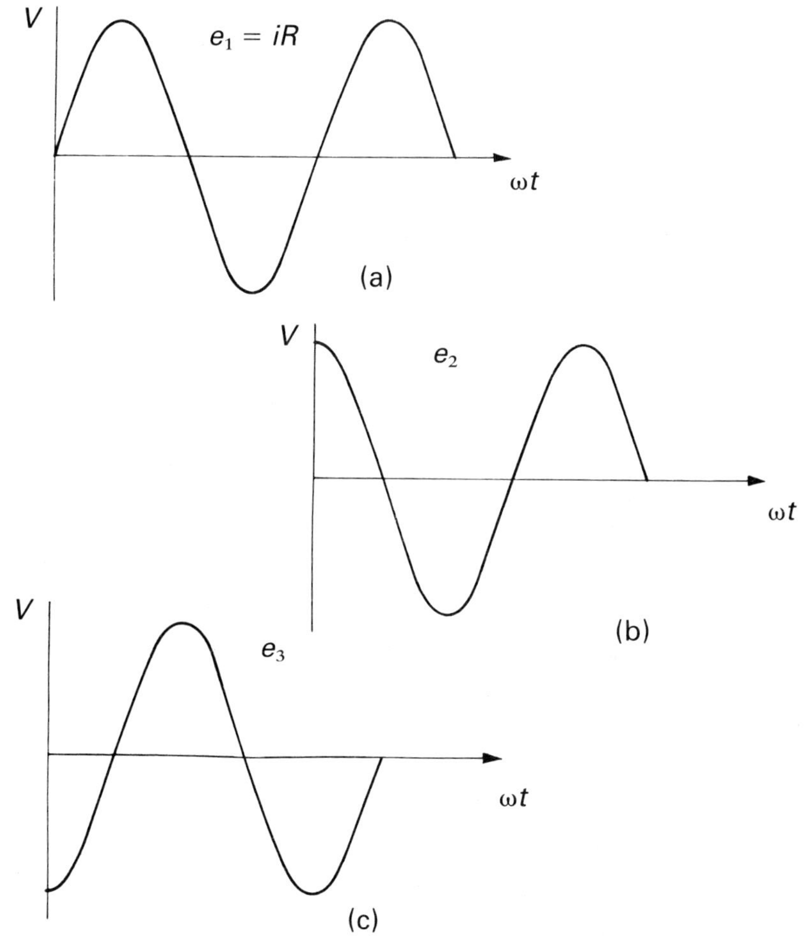

Fig. 1.3 Voltage waveshapes for sinusoidal current in (a) a pure resistance, (b) a pure inductance and (c) a pure capacitance.

there is a time difference between the instants at which the various quantities reach their peak value. This time difference multiplied by the angular frequency ω is called a 'phase' angle. Fig. 1.3 shows that e_2 leads i by $\pi/2$ while e_3 lags i by $\pi/2$. e_1 and i are described as being 'in phase'.

The fact that all the voltage and current variables have the same waveshape when i is sinusoidal is one of the principal reasons why a.c. power networks are driven by sinusoidal voltage generators. The only other function for which the integral and derivative have the same waveshape as the function itself is the exponential function but there are fairly obvious reasons why we prefer sine waves to exponentials in power networks!

If the three circuit elements are now connected in series and carry a current $i = \hat{I} \sin \omega t$ then the driving e.m.f, e, is given by

$$e = e_1 + e_2 + e_3 = \hat{I}R \sin \omega t + \hat{I}\left[\omega L - \frac{1}{\omega C}\right] \cos \omega t \qquad (1.10)$$

The quantities ωL *and* $1/\omega C$ both have the dimension of ohms. They are called reactances. To find the sinusoidal function representing e requires deft trigonometrical manipulation to give

$$e = \hat{I}\left(R^2 + \left(\omega L - \frac{1}{\omega C}\right)^2\right)^{1/2} \sin(\omega t + \theta) \qquad (1.11)$$

where

$$\theta = \tan^{-1} \frac{\left(\omega L - \dfrac{1}{\omega C}\right)}{R}$$

It is clear that when the circuit becomes more complicated than the simple series connection considered here then the trigonometrical manipulation required in the solution process is quite difficult. A more concise and simpler representation is therefore desirable for a.c. variables in order to facilitate the solution process.

1.3 PHASOR REPRESENTATION OF A.C. QUANTITIES

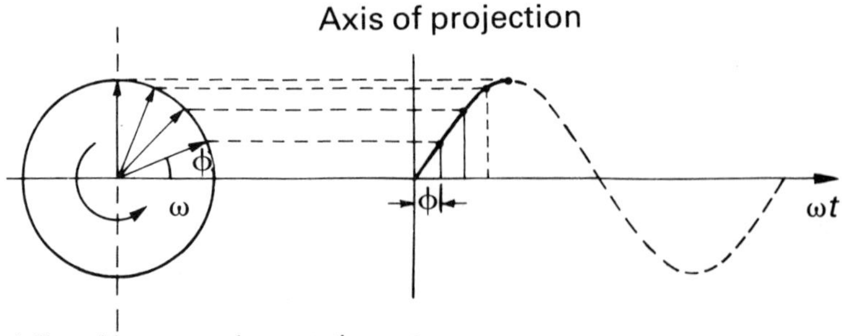

Fig. 1.4 Sinusoid swept out by a rotating vector.

The sinusoidal quantity $\hat{I}\sin\omega t$ can be represented by the projection on a stationary axis of a vector of length $\hat{I}$ rotating at angular frequency ω.

Fig. 1.4 shows several positions of the rotating vector and the corresponding points on the sinusoid represented by the vector projection on a vertical axis. Two sinusoids of the same frequency but with a phase difference between them can be represented by the vertical projections of two such rotating vectors if the angle between the vectors corresponds to the phase difference between the sinusoids.

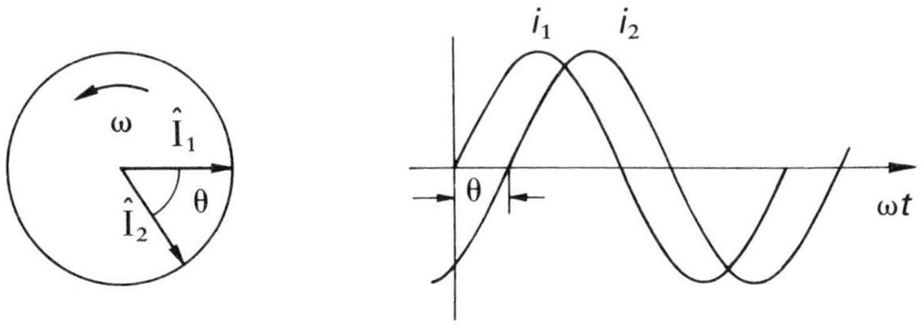

Fig. 1.5 Two sinusoids with phase difference θ and the corresponding rotating vectors.

Fig. 1.5 shows the two sinusoids and their corresponding rotating vectors at the instant $\omega t = 0$. The addition of the two sinusoids i_1 and i_2 can now be achieved by the vector addition of $\hat{\mathbf{I}}_1$ and $\hat{\mathbf{I}}_2$ as shown in Fig. 1.6.

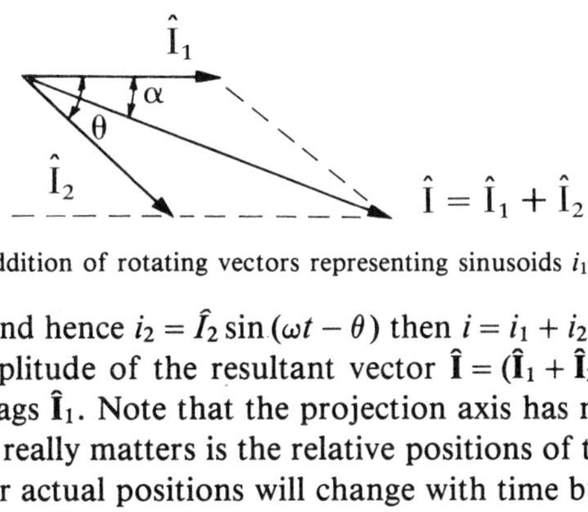

Fig. 1.6 Addition of rotating vectors representing sinusoids i_1 and i_2.

If $i_1 = \hat{I}_1 \sin\omega t$ and hence $i_2 = \hat{I}_2 \sin(\omega t - \theta)$ then $i = i_1 + i_2 = \hat{I}\sin(\omega t - \alpha)$ where $\hat{I}$ is the amplitude of the resultant vector $\hat{\mathbf{I}} = (\hat{\mathbf{I}}_1 + \hat{\mathbf{I}}_2)$ and α is the angle by which $\hat{\mathbf{I}}$ lags $\hat{\mathbf{I}}_1$. Note that the projection axis has not been shown on Fig. 1.6. What really matters is the relative positions of the various vectors. After all their actual positions will change with time but their relative positions remain as in Fig. 1.6. In order to specify the time functions which $\hat{\mathbf{I}}_1$, $\hat{\mathbf{I}}_2$ and $\hat{\mathbf{I}}$ represent it is necessary to know the time function represented by $\bar{\mathbf{I}}_1$ and the phase differences which appear in Fig. 1.6.

A graphical solution, although an improvement on the trigonometrical technique, is no use for an electronic calculator. We require therefore to translate the graphical solution into algebraic terms which can then be dealt with on a calculator. This is achieved by expressing each vector in terms of its horizontal and vertical components and then adding components on each axis to get the components of the resultant vector. In order to clearly distinguish between horizontal and vertical components we can regard the end of a vector as a point in the Argand diagram which has a real part corresponding to the horizontal component and an imaginary part corresponding to the vertical component. For instance, if the vector $\mathbf{I}_1$ corresponds to the point $a + jb$ and $\hat{\mathbf{I}}_2$ correspond to the point $c + jd$ then $(\hat{\mathbf{I}}_1 + \hat{\mathbf{I}}_2)$ corresponds to the point $(a + c) + j(b + d)$. The resultant vector $\hat{\mathbf{I}}$ has amplitude $\sqrt{[(a+c)^2 + (b+d)^2]}$ and makes an angle $\tan^{-1}(b + d)/(a + c)$ with the horizontal axis. Note that we should not confuse components on the real axis of the Argand diagram with the actual value of the sinusoidal quantity which any vector represents at a given instant. The directions of the axes of the Argand diagram are quite arbitrary (apart from being mutually perpendicular). It is the position of the projection axis which dictates the actual value of the sinusoidal quantity represented by the vector.

By way of illustrating this process let us return to the simple example of a circuit comprising R, L and C in series carrying a current $i = \hat{I}\ sin\ \omega t$. From equations (1.1) – (1.10) we have

$$e = \hat{I}R \sin \omega t + \hat{I}\omega L \sin(\omega t + 90) - \frac{\hat{I}}{\omega C}\sin(\omega t + 90) \qquad (1.12)$$

Let us choose the vector $\hat{\mathbf{I}}$ representing the quantity $\hat{I} \sin \omega t$ and let us draw it in the position corresponding to $t = 0$, i.e. in a horizontal position. The

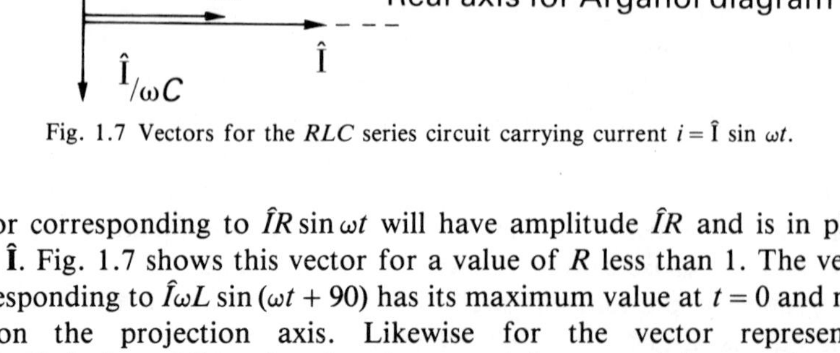

Fig. 1.7 Vectors for the *RLC* series circuit carrying current $i = \hat{\mathrm{I}} \sin \omega t$.

vector corresponding to $\hat{I}R \sin \omega t$ will have amplitude $\hat{I}R$ and is in phase with $\hat{\mathbf{I}}$. Fig. 1.7 shows this vector for a value of R less than 1. The vector corresponding to $\hat{I}\omega L \sin(\omega t + 90)$ has its maximum value at $t = 0$ and must be on the projection axis. Likewise for the vector representing $-(\hat{I}/\omega C) \sin(\omega t + 90)$ but it points downwards because of the negative sign. The vector representing e can now be expressed in terms of its real and imaginary components in the Argand diagram.

$$\hat{\mathbf{E}} = \hat{\mathbf{I}}R + j\hat{\mathbf{I}}\left(\omega L - \frac{1}{\omega C}\right) \tag{1.13}$$

$$= \hat{\mathbf{I}}\left(R + j\left(\omega L - \frac{1}{\omega C}\right)\right)$$

$$= \hat{\mathbf{I}}(R + jX) = \hat{\mathbf{I}}\overline{Z} \tag{1.14}$$

where $X = [\omega L - 1/\omega C]$, is the reactance of the circuit.

The complex quantity $R + jX$ is called the circuit impedance $\overline{Z}$ and is given by the quotient $\hat{\mathbf{E}}/\hat{\mathbf{I}}$ from equation (1.14).

Hitherto we have referred to $\hat{\mathbf{E}}$ and $\hat{\mathbf{I}}$ as vectors but in the preceding sentence we have mentioned the quotient $\hat{\mathbf{E}}/\hat{\mathbf{I}}$. There is no vector operation corresponding to division and in order to evalute $\hat{\mathbf{E}}/\hat{\mathbf{I}} = \overline{Z}, \hat{\mathbf{E}}, \hat{\mathbf{I}}$ and $\overline{Z}$ must all be regarded as complex numbers. To avoid this dilemma $\bar{\hat{E}}$, $\bar{\hat{I}}$ and $\overline{Z}$ will henceforth be called 'phasors' and diagrams such as Fig. 1.7 called 'phasor' diagrams. Equation (1.13) is the phasor form of Kirchhoff's second law and equation (1.14) is the phasor form of Ohm's law. In steady state a.c. problems we can therefore show the variables on a circuit diagram in phasor form and then write the circuit equations directly in phasor form.

Example 1.1

Sketch a phasor diagram for the circuit in Fig. 1.8(a) including the quantity $\bar{\hat{V}}_{AB}$ the potential difference between A and B. If $\bar{\hat{E}}$ is equivalent to a voltage of $100 \sin 314t$ derive the corresponding function of time which $\bar{\hat{V}}_{AB}$ represents.

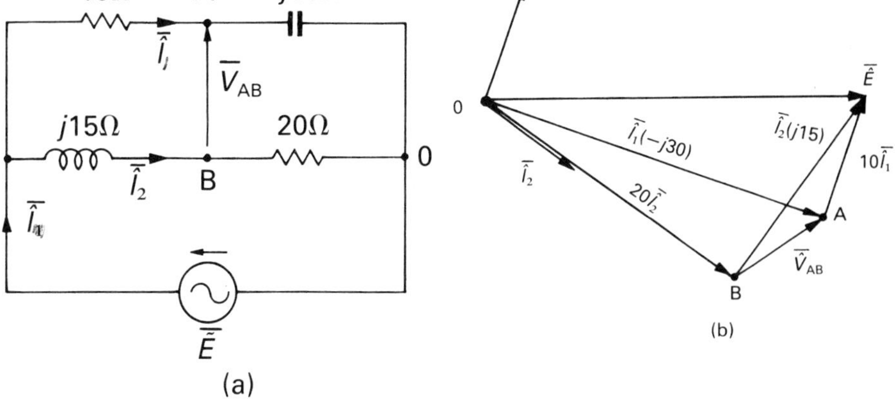

Fig. 1.8 Circuit diagram (a) and corresponding phasor diagram (b).

Note that ωL for the inductance is 15 Ω and $1/\omega C$ for the capacitance is 300 Ω. First, write down the circuit equations in a form which will help to draw a phasor diagram.

$$\bar{\hat{E}} = \bar{\hat{I}}_1 .\ 10 + \bar{\hat{I}}_1 (-j30) \quad (1)$$

and

$$\bar{\hat{E}} = \bar{\hat{I}}_2 (j15) + \bar{\hat{I}}_2 .\ 20 \quad (2)$$

$$\bar{\hat{I}} = \bar{\hat{I}}_1 + \bar{\hat{I}}_2 \quad (3)$$

Let us take $\bar{\hat{E}}$ as reference phasor, i.e. $\bar{\hat{E}} = \hat{E} \angle 0° = \hat{E} + j0$

From equation (1), $\bar{\hat{I}}_1 = \dfrac{\bar{\hat{E}}}{10 - j30} = \dfrac{\bar{\hat{E}}}{31.6} \angle 71.6°$

$\therefore\ \bar{\hat{I}}_1$ leads $\bar{\hat{E}}$ by 71.6° on the phasor diagram.

From equation (2), $\bar{\hat{I}}_2 = \dfrac{\bar{\hat{E}}}{20 + j15} = \dfrac{\bar{\hat{E}}}{25} \angle -36.87°$

$\therefore\ \bar{\hat{I}}_2$ lags $\bar{\hat{E}}$ on the phasor diagram

$\bar{\hat{E}}$, $\bar{\hat{I}}_1$ and $\bar{\hat{I}}_2$ can now be sketched in on Fig 1.8(b).
(Note that the voltage and current scales are different.)

The potential of point A w.r.t. 0 is $\bar{\hat{I}}_1 (-j30)$ and from equation (1)

$$\bar{\hat{I}} (-j30) = \bar{\hat{E}} - \bar{\hat{I}}_1 .10$$

We can now locate point A on the phasor diagram by drawing in the two constituent components of $\bar{\hat{E}}$ from equation (1).

Likewise the potential of B w.r.t. 0 is $20\bar{\hat{I}}_2$ or $\bar{\hat{E}} - \bar{\hat{I}}_2(j15)$ and B can be located on the phasor diagram.

$\bar{\hat{V}}_{AB}$ is the phasor pointing from B to A. The subscript convention will always give the letter to which a phasor points first.

The sketch shows us that $\bar{\hat{V}}_{AB}$ leads $\bar{\hat{E}}$ and is smaller than $\bar{\hat{E}}$.
From the circuit diagrams

$$\bar{\hat{V}}_{AB} = -\ 10\bar{\hat{I}}_1 + j15\bar{\hat{I}}_2$$

or

$$= -\ j30\bar{\hat{I}}_1 - 20I_2$$

and from the first of these expressions

$$\bar{\hat{V}}_{AB} = \frac{-10\bar{\hat{E}}}{31.6} \angle\ 71.6° + \frac{j15\bar{\hat{E}}}{25} \angle\ -36.87°$$

but $\hat{\bar{E}} = 100 \angle 0^\circ$ and $j = 1 \angle 90^\circ$

$$\therefore \hat{\bar{V}}_{AB} = -31.6 \angle 71.6^\circ + 60 \angle 53.13^\circ$$

$$= 31.6 \angle 34.7^\circ$$

$$\underline{v_{ab} = 31.6 \sin(314t + 34.7^\circ)}$$

When circuit elements are in parallel is it more convenient to work with the quantity $\bar{Y} = 1/\bar{Z}$ called the admittance. $\bar{Y}$ is also a complex number and is expressed as $\bar{Y} = G + jB$, where G is called the conductance and B is called the susceptance.

Example 1.2
Calculate the current drawn from the source in Fig. 1.9

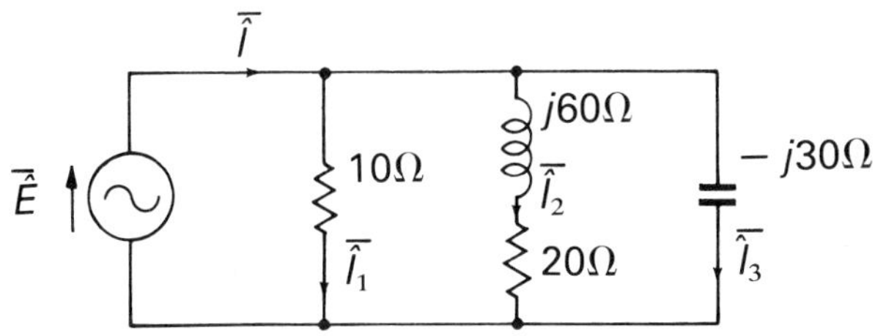

Fig. 1.9 Circuit diagram specifying phasor directions.

$$\hat{\bar{I}} = \hat{\bar{I}}_1 + \hat{\bar{I}}_2 + \hat{\bar{I}}_3 = \frac{\hat{\bar{E}}}{10} + \frac{\hat{\bar{E}}}{20 + j60} + \frac{\hat{\bar{E}}}{-j30}$$

$$= \hat{\bar{E}}\,(\hat{\bar{Y}}_1 + \hat{\bar{Y}}_2 + \hat{\bar{Y}}_3)$$

where

$$\hat{\bar{Y}}_1 = \frac{1}{10} = 0.1 \angle 0^\circ$$

$$\hat{\bar{Y}}_2 = \frac{1}{20 + j60} = 0.014 \angle -63.35^\circ$$

$$\bar{Y}_3 = \frac{1}{-j30} = 0.33 \angle 90^\circ$$

$$\bar{Y}_1 + \bar{Y}_2 + \bar{Y}_3 = 0.108 \angle 10.9^\circ$$

$$\hat{\bar{I}} = \hat{\bar{E}} \angle 0 \times 0.108 \angle 10.9^\circ$$

$$= 100 \times 0.108 \angle 10.9^\circ = \underline{10.8 \angle 10.9^\circ}$$

Hence $i = 10.8 \sin(314t + 10.9^\circ)$
(Note in evaluating i at any instant the quantity $314t$ will be in radians.)

1.4 POWER IN A.C. CIRCUITS

A voltage source $e = \hat{E}\sin(\omega t + \theta)$ delivers a current $i = \hat{I}\sin(\omega t)$ to a load. The power delivered at a given instant will be ei at the instant concerned. Now

$$ei = \hat{E}\hat{I}\sin(\omega t)\sin(\omega t + \theta) = \frac{\hat{E}\hat{I}}{2}\left[\cos\theta - \cos(2\omega t + \theta)\right] \qquad (1.15)$$

The terms in the bracket in equation (1.15) indicate that the power flow consists of two components. The first term $(\hat{E}\hat{I}/2)\cos\theta$ is independent of time while the second term $(\hat{E}\hat{I}/2)\cos(2\omega t + \theta)$ oscillates at twice the power frequency. The average power given by

$$P = \frac{1}{2\pi}\int_0^{2\pi} ei \, \mathrm{d}(\omega t)$$

is therefore $(\hat{E}\hat{I}/2)\cos\theta$. This expression is usually stated as $P = EI\cos\theta$, where E and I are the root mean square values of e and i respectively. (The r.m.s. value of a sinusoidal quantity is its amplitude divided by $\sqrt{2}$.)

1.4.1 Watts and Vars

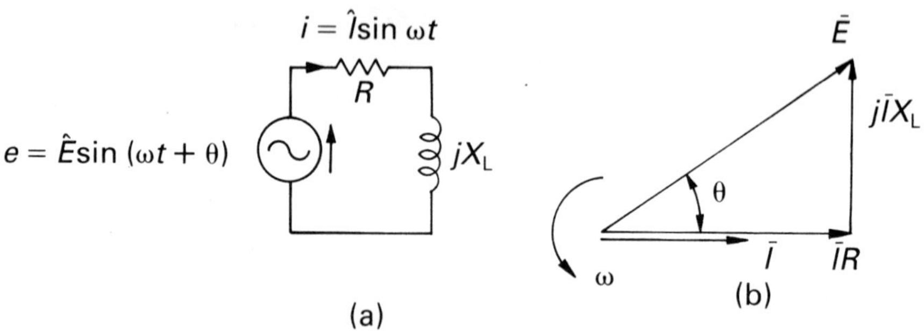

Fig. 1.10 Circuit diagram (a) and corresponding phasor diagram (b).

Figure 1.10(a) shows a voltage source supplying a resistance and inductance in series. Figure 1.10(b) shows the corresponding phasor diagram with all phasors scaled by $\sqrt{2}$ to give r.m.s. values. This will henceforth be the practice with all phasor diagrams. Expressed as complex numbers, $\bar{I} = I + j0 = I \angle 0°$ in polar form and $\bar{E} = E\cos\theta + jE\sin\theta = E \angle \theta$ in polar form. The product $\bar{E}\bar{I} = EI \angle \theta$ can be expanded to give the complex number $(EI\cos\theta + jEI\sin\theta)$. The first term of this expansion is identified as the average power and is measured in watts. We might be forgiven for thinking that the second term is an imaginary quantity as a result of the prefix j. Power engineers, however, find this term of sufficient importance to give it the special name of Volt Amperes Reactive, Vars for short.

Dimensionally there is no difference between watts and vars, the different names arising from their different physical significance.

Consider the 'instantaneous watts' in each element of Fig. 1.10(a). To do this we need to know the voltage and current for each element. The current in both cases is $\hat{I}\sin\omega t$ and from the phasor diagram of Fig. 1.10(b) we can see that the voltage across R is $\hat{E}\cos\theta\sin\omega t$ (in phase with i) and the voltage across L is $\hat{E}\sin\theta\sin(\omega t+90)$ [90° ahead of i].

The total ei product, therefore, maintaining the identity of each element, is

$$ei = \hat{E}\cos\theta\sin\omega t.\ \hat{I}\sin\omega t + \hat{E}\sin\theta\sin(\omega t+90).\ \hat{I}\sin\omega t$$

$$= \hat{E}\hat{I}\cos\theta\sin^2\omega t + \hat{E}\hat{I}\sin\theta\cos\omega t\sin\omega t \qquad (1.16)$$

(The expressions in equations (1.15) and (1.16) are equivalent) In terms of r.m.s. quantities the 'instantaneous' watts is given by

$$ei = EI\cos\theta(1-\cos 2\omega t) + EI\sin\theta\sin 2\omega t \qquad (1.17)$$

The first term on the right-hand side is due to the resistor and the second term due to the inductance.

Fig. 1.11 shows sketches of these two terms with e and i shown for comparison.

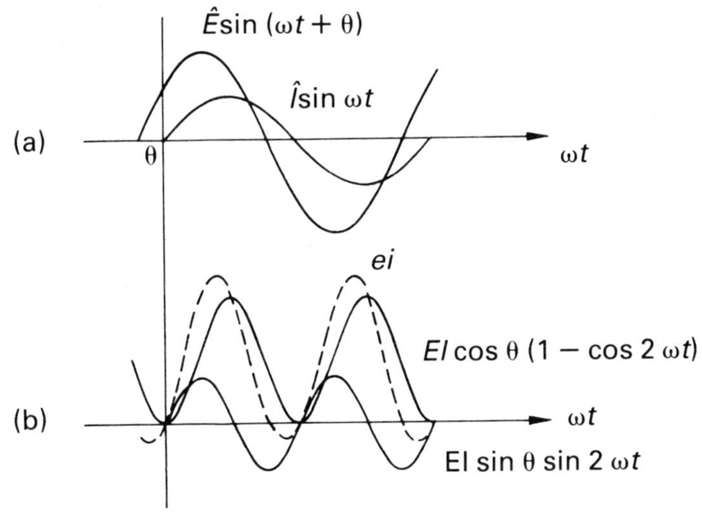

Fig. 1.11 Voltage and current waveforms (a) and corresponding voltamperes, watts and vars (b).

The component due to R can be seen to have an average value of $EI\cos\theta$ while the component due to L has zero average value. Pulsating components at twice the source frequency are present in both R and L and the amplitude of the pulsating component in L is $EI \sin\theta$. $EI \sin\theta$ is therefore a measure of the rate at which energy is exchanged between the source and the energy storage element L. The presence of an energy storage element in the circuit affects the value of the peak *ei* product as can be seen from Fig. 1.11(b). This value is in excess of the peak value for R alone.

The vars in circuit 1.10(a) are therefore seen to be the amplitude of oscillating watts in the energy storage element. The energy storage element in the case considered was L but in general it may be either L or C or a combination of both. Such elements give rise to a reactive component in the impedance expression and vars are therefore required whenever the circuit impedance contains reactance.

$EI \cos\theta$ although strictly speaking the 'average' power is invariably referred to as 'power' and the reader should therefore bear in mind that this 'power' is not constant even for a resistive element.

In the expression for power, $EI \cos\theta$, the term $\cos\theta$ is referred to as the 'power factor'. For a purely resistive circuit $\cos\theta = 1$.

1.4.2 The use of $\bar{E}\bar{I}^*$ or $\bar{E}^*\bar{I}$

In the previous section the phasor multiplication $\bar{E}.\bar{I}$ yielded the expression $EI \angle \theta = EI\cos\theta + jEI\sin\theta$. $\bar{I}$ was the reference phasor $I \angle 0$ and as a result the j term is positive if $\bar{E}$ leads $\bar{I}$ by an angle θ. If $\bar{E}$ had been the reference phasor then $\bar{I}$ would have been $I \angle -\theta$ and $\bar{E}.\bar{I}$ would have given a negative j term. The load impedance $(R + jX)$ has not changed and yet the sign of the vars appears to depend on our choice of reference phasor! The situation becomes further confused if neither $\bar{E}$ nor $\bar{I}$ is the reference phasor, i.e. $\bar{E} = E \angle \alpha$, $\bar{I} = I \angle \beta$. The phasor product $\bar{E}.\bar{I}$ gives $EI \angle \alpha + \beta = EI\cos(\alpha + \beta) + jEI\sin(\alpha + \beta)$. Neither of the two latter terms is in terms of the angle $\theta = (\alpha - \beta)$, the angle between $\bar{E}$ and $\bar{I}$. To rectify this situation we must consider the phasor product $\bar{E}.\bar{I}^* = E\angle\alpha . I\angle -\beta = EI\cos(\alpha - \beta) + jEI\sin(\alpha - \beta) = EI\cos\theta + jEI\sin\theta$. ($\bar{I}^*$ is called the complex conjugate of I.) Provided $\alpha > \beta$, i.e. the current lags the voltage, then θ is positive and the vars are positive. This is the usual convention for vars wherein a load which draws a lagging current absorbs positive vars. (The use of the phasor product $\bar{E}^*.\bar{I}$ will give the same term for the power but will change the sign of the vars since $\sin(-\theta) = -\sin\theta$.) With the above convention for positive vars it is therefore simpler to use the expression $\bar{E}.\bar{I}^*$ which automatically gives the correct sign for vars and avoids any problems as to whether $\bar{E}$ or $\bar{I}$ or neither is the reference phasor. The analysis carried out in Section 1.4.1 is still valid since $\bar{E}.\bar{I} = \bar{E}.\bar{I}^*$ when $\bar{I} = I \angle 0$.

When $\bar{E}$ and $\bar{I}$ are expressed as complex numbers in the $a + jb$ form the watts and vars can be obtained directly from the expression $\bar{E}\bar{I}^*$ or $\bar{E}^*\bar{I}$.

If $\bar{E} = (a + jb)$ and $\bar{I} = (c + jd)$ then $\bar{E}.\bar{I}^* = (a + jb)(c - jd) = (ac + bd) + j(bc - ad)$. But average power is $EI\cos\theta$ and vars is $EI\sin\theta$, where θ is the angle between $\bar{E}$ and $\bar{I}$. Let us evaluate the expressions $EI\cos\theta$ and $EI\sin\theta$ in terms of a, b, c and d.

$$\bar{E} = a + jb = \sqrt{(a^2 + b^2)} \ \angle\ \alpha,\ \alpha = \tan^{-1}\frac{b}{a}$$

$$\bar{I} = c + jd = \sqrt{(c^2 + d^2)} \ \angle\ \beta,\ \beta = \tan^{-1}\frac{d}{c}$$

$$\theta = \alpha - \beta$$

$$EI\cos\theta = \sqrt{(a^2 + b^2)}\sqrt{(c^2 + d^2)}\cos(\alpha - \beta)$$

$$\cos(\alpha - \beta) = \cos\alpha\cos\beta + \sin\alpha\sin\beta$$

$$= \frac{a}{\sqrt{(a^2 + b^2)}} \cdot \frac{c}{\sqrt{(c^2 + d^2)}} + \frac{b}{\sqrt{(a^2 + b^2)}} \cdot \frac{d}{(c^2 + d^2)}$$

$$= \frac{ac + bd}{\sqrt{[(a^2 + b^2)(c^2 + d^2)]}}$$

hence $EI\cos\theta = ac + bd =$ real part of $\bar{E}\bar{I}^*$.

By a similar process $EI\sin\theta = bc - ad =$ imaginary part of $\bar{E}\bar{I}^*$

If we use $\bar{E}^*\bar{I}$ then $EI\cos\theta$ is still equal to the real part but $\bar{E}\bar{I}\sin\theta$ is now equal to—(the imaginary part).

1.5 POWER FACTOR CORRECTION

When a voltage source supplies power to a load impedance $\bar{Z}_L$, say, via an interconnector of non-zero impedance then both a voltage drop and ohmic losses will occur in the interconnector. If the voltage source represents the output terminals of the power supply company and the load impedance $\bar{Z}_L$ represents the power consumer then the consumer wishes to be charged only for the power consumed on his premises and furthermore he wants that power delivered within a few per cent difference of a stated nominal voltage irrespective of variations in $\bar{Z}_L$. The power company on the other hand has to supply both the power drawn by the consumer and the losses in the interconnector. In order to minimise the latter the power company encourages consumers to draw their power at minimum current. The 'encouragement' takes the form of charging large consumers for vars as well as watts.

Consider the following simple example.

Example 1.3

Fig. 1.12 shows a source $\bar{E}$ supplying a load $\bar{Z}_L$ via an interconnector of impedance $(0.1 + j0.5)\Omega$. The interconnector impedance represents the combined impedance of both the go and return conductors.

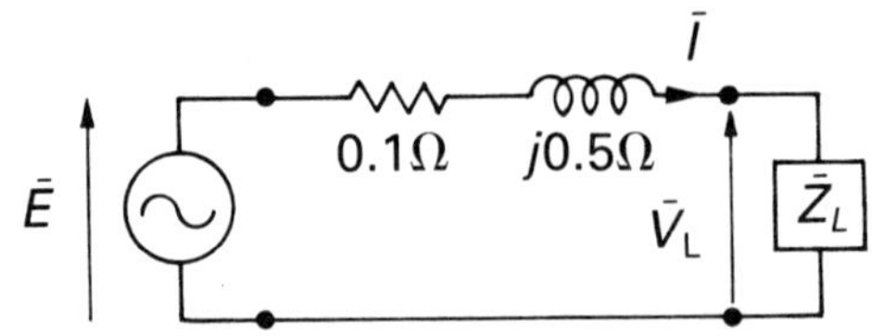

Fig. 1.12 Source ($\bar{E}$) supplying load ($\bar{Z}_L$) via an interconnector.

If E is 240 V and $\bar{Z}_L = (8 + j6)\Omega$ then

$$\bar{I} = \frac{240 \angle 0}{(8.1 + j6.5)} = 23.11 \angle -38.75^\circ$$

We have taken $\bar{E}$ as reference phasor and all phase angles will therefore be given with reference to $\bar{E}$.

The load voltage $\bar{V}_L$ is given by

$$\bar{V}_L = \bar{I}\bar{Z}_L = 23.11 \angle -38.75^\circ \times 10 \angle 36.87^\circ$$
$$= 231.1 \angle -1.88^\circ$$

The voltage drop in the interconnector is $240 \angle 0^\circ - 231.1 \angle -1.88^\circ = 11.8 \angle 40^\circ$ and there is a 8.9 V difference in voltage amplitude. The power loss is $I^2 \times 0.1 = 53.4$ watts. The consumer draws power of $I^2 \times 8 = 4.27$ kW, and vars of $I^2 \times 6 = 3.2$ kVars and will be charged at a different rate for each. The consumer can reduce his var consumption by the simple expedient of connecting a capacitor in parallel with $\bar{Z}_L$ such that the parallel combination of $\bar{Z}_L$ and C looks like a resistance.

To simplify the algebra let us keep $\bar{V}_L$ at its previous value and with reference to Fig. 1.13 we can see that $\bar{I}_c$ has to cancel the out of phase component of $\bar{I}$ to give a net current in the interconnector of $\bar{I} + \bar{I}_c$ in phase with $\bar{V}_L$. This means that the power factor for the combined load ($\bar{Z}_L$ and C) is unity.

The component of $\bar{I}$ at right angles to $\bar{V}_L$ is $I \sin 36.87^\circ = 13.87$ A

The capacitor C draws a current $\bar{I}_c$ of amplitude 13.87 A leading $\bar{V}_L$ by 90° and hence cancelling the lagging component of $\bar{I}$. The net interconnector current

$$\bar{I} + \bar{I}_c = \bar{I} \cos 36.87^\circ = 18.49 \text{ A}$$

The new applied volts $\bar{E}$ is given by

$$\bar{E} = \bar{V}_L + (\bar{I} + \bar{I}_c)\,(0.1 + j0.5)$$

and taking $\bar{V}_L$ as our reference phasor for this calculation

$$\bar{E} = 231.1 \angle 0^\circ + 18.49 \angle 0^\circ \times 0.51 \angle 78.69^\circ = 233.1 \angle 2.27^\circ$$

The voltage drop in the interconnector is $233.1 \angle 2.27^\circ - 231.1 \angle 0^\circ = 9.41 \angle 78.87^\circ$ and there is 2 V difference in voltage amplitude. The loss is $18.49^2 \times 0.1 = 34.2$ watts. All these figures are less than in the previous case.

The catch as far as the consumer is concerned is that he will have to buy a capacitor C capable of handling $V_L.I_c$ vars. This cost has then to be offset against the avoidance of recurrent expenditure on vars drawn from the power company.

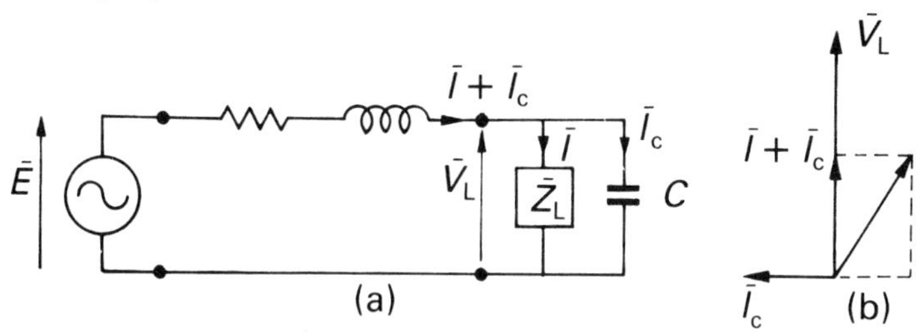

Fig. 1.13 Circuit (a) and phasor diagram (b) for load with power factor correction.

Domestic consumers already draw power at substantially unity power factor and no measurement or charge for vars is therefore made. Industrial consumers on the other hand tend to draw power at lagging power factors and are charged for vars as well as watts.

1.6 SPECIFICATION OF LOADS

In Example 1.3 the load was specified as an impedance. It is common practice in power systems which run within a few per cent of a nominal voltage to specify loads in terms of the nominal voltage, the power they consume at this voltage and the power factor at which they draw this power. On the basis that the nominal system voltage in the last example was 240 V then the load (Z_L) would have been specified as 240 V, 4.27 kW at 0.8 p.f. lagging.

Example 1.4

A load draws 4 MW at 0.8 p.f. lagging and at a voltage of 11 kV. It draws power from a supply via an interconnector of impedance $(0.1 + j0.5)\Omega$. Calculate the supply voltage magnitude.

The circuit diagram is as in Fig. 1.12 with V_L equal to 11kV and Z_L replaced by the 4 MW load. First of all we calculate the interconnector current I.

$$V_L.\ I.\cos\theta = 4 \times 10^6\ \text{W}$$

$$\therefore I = \frac{4 \times 10^6}{11 \times 10^3 \times 0.8} = 454.54\ \text{A}$$

If $\bar{V}_L$ is taken as reference phasor then $\bar{I} = 454.54 \angle -36.87°$. We can now calculate $\bar{E}$ as before from

$$\bar{E} = 11 \times 10^3 \angle 0° + 454.54 \angle -36.87° \times 0.51 \angle 78.69°$$

or we can make use of the theorem of conservation of watts and vars. This theorem comes from the conservation of power applied to the instantaneous power *ei*. In this case it means that the watts from the supply must be the sum of the watts consumed in the interconnector and the load and likewise for the vars.

$$\text{Interconnector watts and vars} = I^2 \times 0.1 + jI^2 \times 0.5$$
$$= 20.66 \text{ kW} + j103.3 \text{ kVars}$$
$$\text{The load watts and vars} = 4 \times 10^6 + j(3 \times 10^6)$$
$$\therefore \text{ Supply watts and vars} = 4.02 \text{ MW} + j3.103 \text{ MVars}$$
$$\text{Hence supply volt-amperes} = \sqrt{(\text{watts}^2 + \text{vars}^2)}$$
$$= 5.079 \text{ MVA}$$
$$\text{But supply volt-amperes} = E.I.$$

hence
$$E = \frac{5.079 \times 10^6}{454.54}$$
$$= \underline{11.173 \text{ kV}}$$

1.7 TWO-PHASE SYSTEMS

If we have two loads of magnitude $\bar{Z}_L$ we could connect them in parallel and supply them from a single phase source $\bar{E}$. The interconnector cable would have to be capable of carrying a current of magnitude $2E/Z_L$. On the other hand if we have a two-phase supply, the loads can be connected as in Fig. 1.14(a). Each load carries the same current as before but there is a

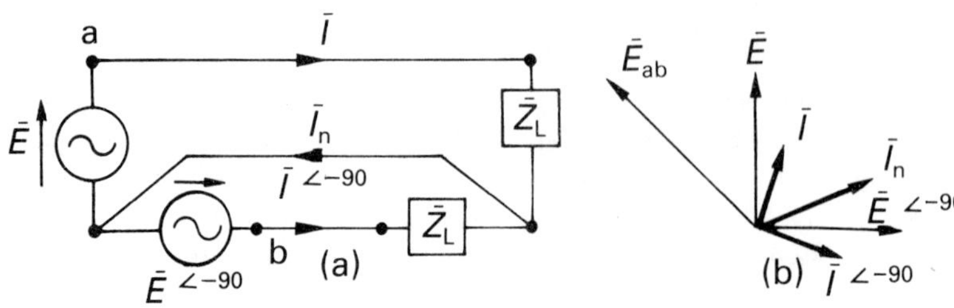

Fig. 1.14 Circuit diagram (a) and phasor diagram (b) for a balanced two-phase system.

phase difference between the two currents which results in the central return conductor carrying $I_n = \sqrt{2}\, I$ and not $2I$ as in the single phase case. This system will therefore save on interconnector copper but against this must be weighed the more complicated source structure. The voltage between the two points a and b in Fig. 1.14(a) is not $2E$ but $\sqrt{2}E$ as can be seen from Fig. 1.14(b). This latter voltage is called a 'line' voltage to distinguish it from the phase voltage E. A line voltage is measured between two 'lines' whereas a phase voltage is measured between a 'line' and the neutral return conductor.

1.8 THREE-PHASE SYSTEMS

1.8.1 General considerations

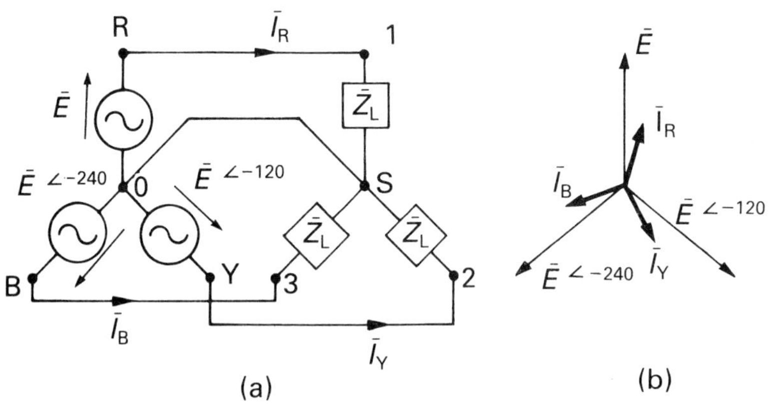

Fig. 1.15 Circuit diagram (a) and phasor diagram (b) for a balanced three-phase system.

Again if we have three loads of $\bar{Z}_L$ they could be connected in parallel and fed from a single phase source $\bar{E}$. The interconnector would be required to carry a current of magnitude $3E/Z_L$. Alternatively if a three-phase source is available the interconnection shown in Fig. 1.15(a) could be used. Three-phase supply terminals are commonly labelled RYB to correspond to the red, yellow and blue cable colours used in three-phase wiring. Each wire in the interconnector other than OS carries a current of magnitude E/Z_L and the neutral return OS carries $\bar{I}_n = \bar{I}_1 + \bar{I}_2 + \bar{I}_3$ which from the phasor diagram of Fig. 1.15(b) can be seen to be zero. The points O and S are therefore at the same potential and the wire OS can be dispensed with. Again there is a saving in interconnector copper to be weighed against a more complicated source structure. But there is one more very important factor for a three-phase system which is not present in the single-phase system.

Consider the total instantaneous power supplied to the load.

For phase 1. $ei = \hat{E}\sin\omega t\ \hat{I}\sin(\omega t - \theta)$

For phase 2. $ei = \hat{E}\sin(\omega t - 120)\hat{I}\sin(\omega t - 120 - \theta)$

For phase 3. $ei = \hat{E}\sin(\omega t - 240)\hat{I}\sin(\omega t - 240 - \theta)$

It is left as an exercise for the reader to show that the sum of these three terms gives $(3\hat{E}\hat{I}/2)\cos\theta = 3EIcos\theta$. The total instantaneous power is independent of time and this means there are no power pulsations in say large three-phase machines where a pulsating torque could lead to serious vibration problems. Since there is no term in $EI\sin\theta$ we might be tempted to suppose that there is no need to supply vars to the load. Unfortunately in each phase there is still an exchange of energy between the source and the reactive element and the load would still be specified in terms of the three-phase power ($3EI\cos\theta$) and the power factor ($\cos\theta$).

The property of constant instantaneous power is a property of any symmetrical n-phase system, e.g. $n = 4$ and a source with four equal e.m.f's all at 90° to each other. It is also true that the saving in copper continues to increase as n increases. However, the increasing complexity of the source, transmission lines and other system elements indicate that $n = 3$ is the optimum solution.

Small loads up to say 30 kW where the power pulsations do not lead to serious vibrations will be supplied on a single-phase basis and larger loads will be supplied on a three-phase basis. Two-phase systems have almost all been replaced by three-phase systems and are not worthy of any further consideration.

1.8.2 Star and delta configurations

Given a three-phase voltage source, i.e. three voltages of equal magnitude with 120° phase difference between any two phasors there are two possible configurations in which the source may be used to supply a load. Fig. 1.16 shows these two configurations and their corresponding phasor diagrams.

The voltmeter shown dotted in Fig. 1.16(a) will read a voltage equivalent to the e.m.f. in the source between points Y and R, i.e. $|\bar{E}_{RY}|$. From the geometry of the phasor diagram of Fig. 1.16(a) $|\bar{E}_{RY}| = \sqrt{3}\,|\bar{E}|$. In a star connected system the voltage between any two lines, RY, YB or BR is $\sqrt{3}$ times larger than the phase voltage $|\bar{E}|$. In the delta connected system of Fig. 1.16(b) the voltmeter between lines R and Y clearly reads $|\bar{E}|$ and the voltage between any two lines will have magnitude E. The line voltage in such a system is therefore of equal magnitude to the phase voltage.

The three interconnector lines R, Y and B will connect to a three-phase load. The load too will be connected in a star or delta configuration. It is simpler in the first instance to consider systems for which the source and load have the same configuration.

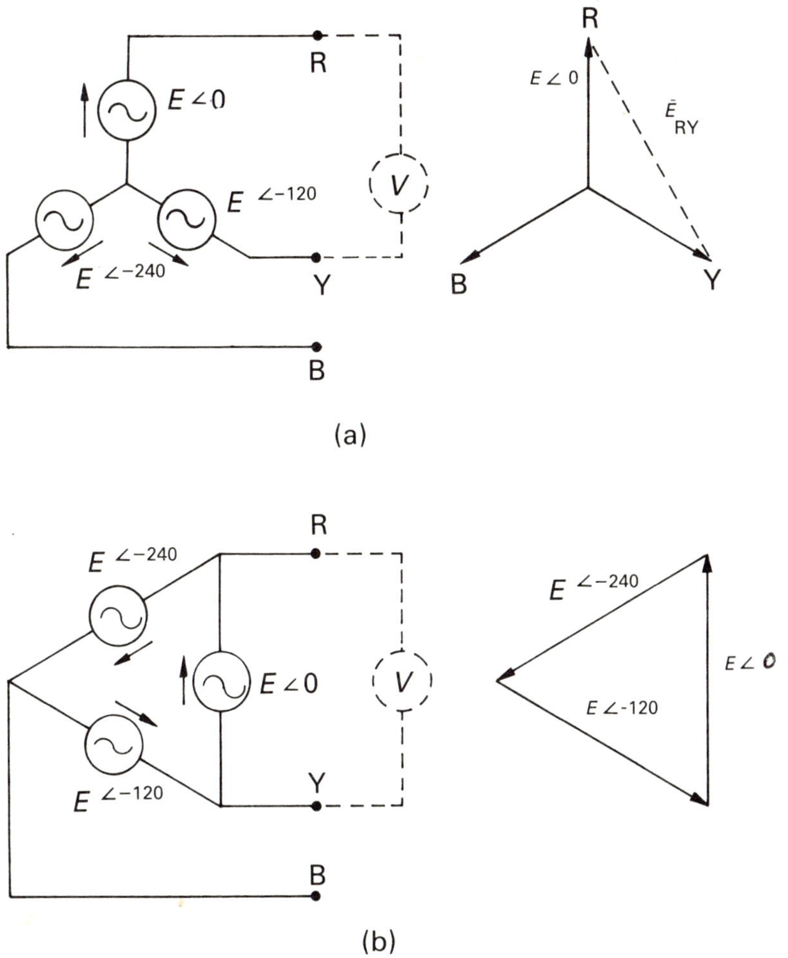

Fig. 1.16 Measurement of line voltage in a balanced three-phase (a) star connected system and (b) delta connected system.

Fig. 1.15 shows the connection for a star connected system. The line voltage as shown in Fig. 1.16(a) will be $\sqrt{3}$ times the phase voltage. The current flowing in any phase of the load ($\bar{Z}_L$) is equal to the current flowing in the line connected to that phase so that the line and phase currents are equal in magnitude. If we use the subscript p for phase quantities and L for line quantities then for a star connected system

(1) $V_L = \sqrt{3} V_p$
(2) $I_L = I_p$
(3) Total power $= 3 V_p I_p \cos\theta = \sqrt{3} V_L I_L \cos\theta$

Fig. 1.17(a) shows a delta connected system.

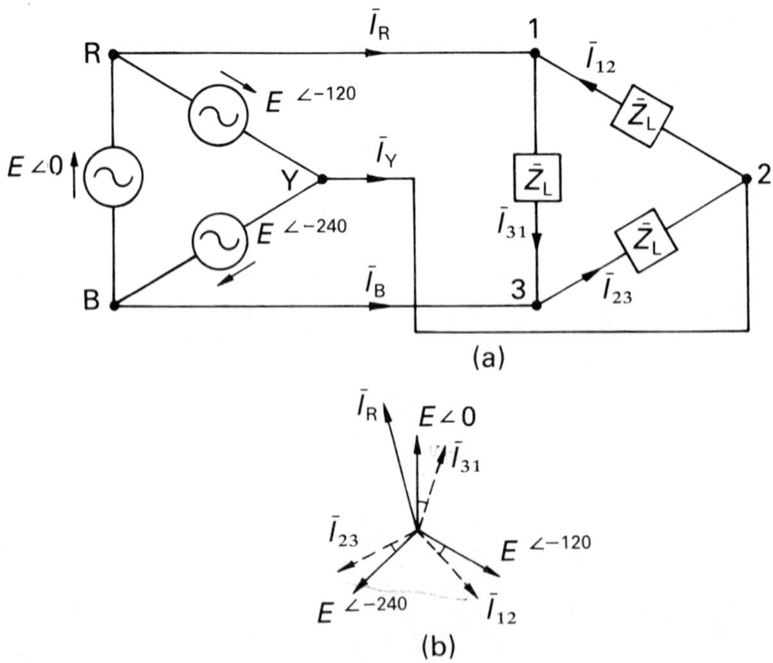

Fig. 1.17 Currents in a balanced three-phase delta connected system (a) circuit diagram, (b) phasor diagram.

The line currents $\bar{I}_R$,$\bar{I}_Y$ and $\bar{I}_B$ cannot be found directly but must be calculated from the load phase currents $\bar{I}_{12}$, $\bar{I}_{23}$ and $\bar{I}_{31}$. To find the phase current $\bar{I}_{12}$, for instance, consider the loop RY21R. Assuming that the interconnector impedance is negligible then

$$E \angle -120^\circ = \bar{I}_{12}\,\bar{Z}_L$$

$$\therefore\ \bar{I}_{12} = \frac{E \angle -120^\circ}{Z_L \angle \theta^\circ}$$

$$\text{Similarly } \bar{I}_{23} = \frac{E \angle -240^\circ}{Z_L \angle \theta^\circ},\quad \bar{I}_{31} = \frac{E \angle 0^\circ}{Z_L \angle \theta^\circ}$$

Kirchhoff's first law at node 1 gives

$$\bar{I}_R = \bar{I}_{31} - \bar{I}_{12}$$

and from the geometry of the phasor diagram in Fig. 1.17(b) $|\bar{I}_R| = \sqrt{3}\,|\bar{I}_{31}|$. $\bar{I}_B$ and $\bar{I}_Y$ can be found by summing currents at nodes 2 and 3 respectively. They have the same magnitude as $\bar{I}_R$ but will be lagging $\bar{I}_R$ by 120° and 240° respectively.

A voltmeter connected between any two lines will read a voltage of magnitude E.

Hence for a delta connected system

(1) $V_L = V_p$
(2) $I_L = \sqrt{3} I_p$
(3) Total power $= 3 V_p I_p \cos\theta = \sqrt{3} V_L I_L \cos\theta$

We can conclude therefore that for a balanced three-phase load the total power is always given by $\sqrt{3} V_L I_L \cos\theta$ for both star and delta systems. Likewise the total VA is $\sqrt{3}\ V_L I_L$ and the total vars is $\sqrt{3}\ V_L I_L \sin\theta$.

It is also clear from the phasor diagrams of Figs. 1.15(b) and 1.17(b) that each set of phasors, the phase voltages, the line voltages, the phase currents, the line currents, is a symmetrical three-phase set, i.e. three phasors of equal magnitude all at 120° to each other. There is never any need to calculate any more than one quantity in each set. The remaining two can then be inferred from symmetry.

Example 1.5

For the circuit of Fig. 1.15(a) with $E = 240$ V and $\bar{Z}_L = (8 + j6)\,\Omega$ calculate the phase currents, the line voltages and total power.

Let the reference phasor be the red-phase (R) e.m.f. Since the load and supply star points are at the same potential then from the red-phase

$$\bar{I}_R = \frac{E \angle 0^\circ}{8 + j6} = \underline{24 \angle -36.87^\circ \text{ amperes}}$$

We can now write

$$\bar{I}_Y = 24 \angle -36.87^\circ - 120^\circ$$

$$= \underline{24 \angle -156.87^\circ \text{ amperes}}$$

and

$$\bar{I}_B = 24 \angle -36.87^\circ - 240^\circ$$

$$= \underline{24 \angle -276.87^\circ \text{ amperes}}$$

We must exercise care in choosing which current is 120° behind $\bar{I}_R$ and which is 240°. Sketching a phasor diagram helps to get this right but in any case the currents will follow the same sequence as the supply e.m.f's i.e. RYB.

The voltage between lines R and Y, Fig. 1.15(a), is given by

$$\bar{V}_{RY} = -E \angle -120^\circ + E \angle 0$$

$$= \sqrt{3}E \angle 30^\circ = \underline{415.7 \angle 30^\circ \text{ V}}$$

The other voltages can now be written down as

$$\bar{V}_{YB} = 415.7 \angle -90^\circ \text{ V}$$

$$\bar{V}_{BR} = 415.7 \angle -210^\circ \text{ V}$$

$$\begin{aligned}\text{The total power } &= \sqrt{3} V_L I_L \cos\theta \\ &= \sqrt{3} \times 415.7 \times 24 \times 0.8 \\ &= \underline{13.82 \text{ kW}}\end{aligned}$$

(For the unconvinced check that this is the same as $3V_p I_p \cos\theta$ or $3I_p^2 \times 8$)

1.8.3 Specification of three-phase supply and load

As in the single phase case a balanced three-phase load would be specified in terms of the nominal line voltage, the total power and the load power factor. The load in Example 1.5 would be specified as 415 V (nominal) 13.8 kW at 0.8 p.f. lagging. The load configuration would only be necessary if there was need to know the load phase currents.

The supply would be specified in terms of its line voltage, frequency, and phase sequence, i.e. 415.7 V, 50 Hz and sequence RYB. Again there is no need to know the phase configuration unless supply phase currents are required. A further example will show how to use information presented in this format.

Example 1.6

A 415 V, 50 Hz, three-phase supply of phase sequence RYB is connected to a 415 V, 13.8 kW, 0.8 p.f. lagging, delta connected load. Calculate the line current and the load phase current.

$$\sqrt{3} V_L I_L \cos\theta = 13.8 \times 10^3 \text{ W}$$

$$\therefore\ I_L = \frac{13.8 \times 10^3}{\sqrt{3} \times 415 \times 0.8} = \underline{24 \text{ amperes}}$$

$$\begin{aligned}\text{The load phase current } I_p &= I_L/\sqrt{3} \\ &= \underline{13.86 \text{ amperes}}\end{aligned}$$

$\bar{I}_p$ will lag the voltage across that phase of the load by $\cos^{-1} 0.8$. The line currents and phase currents will each form a three-phase set of currents with a 30° displacement between the two sets as shown in Fig. 1.17(b).

1.8.4 Unbalanced three-phase loads

A load is described as unbalanced when the impedances in each of its three branches are not all equal. Such a load cannot be adequately described in terms of its total power but must have each branch (phase) of the load separately specified. Neither will the line and phase currents form symmetrical three phase sets and separate solution is required for each current. The added complexity of the problem is best illustrated by an example.

Example 1.7

A 400 V, 50 Hz, three-phase supply of phase sequence RYB is connected to an unbalanced delta load as shown in Fig. 1.18. Calculate the line currents and total power.

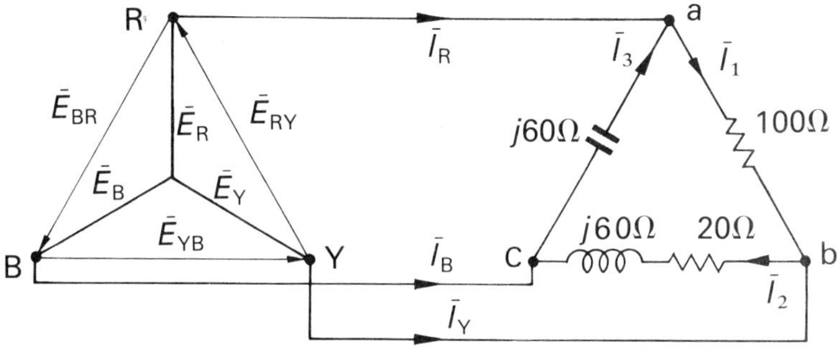

Fig. 1.18 Combined circuit diagram and voltage phasor diagram for unbalanced delta connected load.

The supply is shown in terms of the phasors which give rise to the line voltages. There is no need to know the supply configuration. We must first of all calculate the branch currents in the load.

Loop *abYRa*

$$\bar{E}_{\mathrm{RY}} = \bar{I}_1.100 \angle 0^\circ \text{ V}$$

Take $\bar{E}_{\mathrm{RY}}$ as reference phasor and hence

$$\bar{I}_1 = \frac{400 \angle 0^\circ}{100 \angle 0^\circ} = \underline{4 \angle 0^\circ \text{ amperes}}$$

Loop *bcBYb*

$$\bar{E}_{\mathrm{YB}} = \bar{I}_2(20 + j60) \text{ V}$$

$$\therefore \bar{I}_2 = \frac{400 \angle -120^\circ}{20 + j60} = \underline{6.32 \angle 168.4^\circ \text{ amperes}}$$

The only difficulty here is to choose the correct angle for $\bar{E}_{\mathrm{YB}}$. Sketching the supply phasor diagram is of great assistance.

Loop *caRBc*

$$\bar{E}_{\mathrm{BR}} = \bar{I}_3.106 \angle -90^\circ \text{ V}$$

$$\therefore \bar{I}_3 = \frac{400 \angle 120^\circ}{106 \angle -90^\circ} = \underline{3.77 \angle 210^\circ \text{ amperes}}$$

The line currents can now be found by summing currents at the nodes a, b, c.

$$\left.\begin{aligned}\bar{I}_R &= \bar{I}_1 - \bar{I}_3 = 7.5 \angle 14.5^\circ \text{ amperes}\\ \bar{I}_Y &= \bar{I}_2 - \bar{I}_1 = 10.27 \angle 172.9^\circ \text{ amperes}\\ \bar{I}_B &= \bar{I}_3 - \bar{I}_2 = 4.3 \angle -47.2^\circ \text{ amperes}\end{aligned}\right\} \quad \begin{array}{l}\text{Note}\\ \Sigma \bar{I} = 0\\ \text{for line currents}\end{array}$$

The total power consumed by the load is

$$100\, I_1^2 + 20\, I_2^2 = \underline{2399 \text{ W}}$$

* * *

The unbalanced delta load can still be dealt with phase by phase. The unbalanced star load presents a more difficult problem.

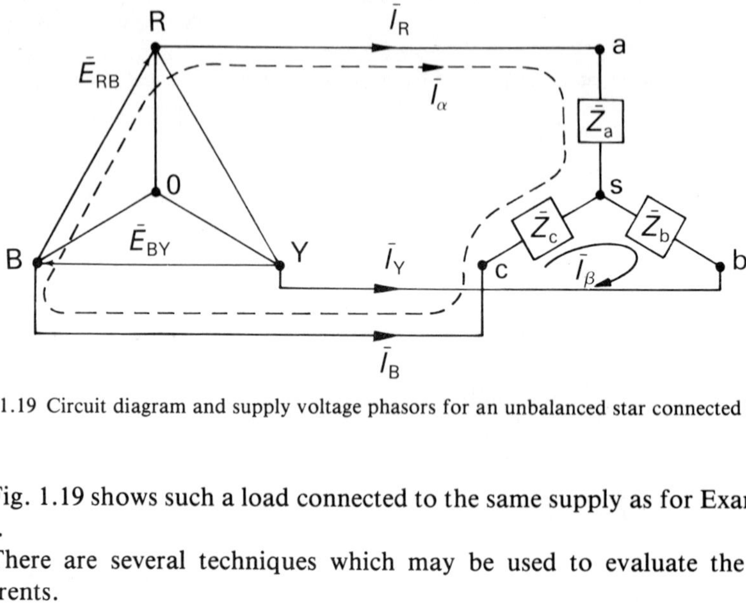

Fig. 1.19 Circuit diagram and supply voltage phasors for an unbalanced star connected load.

Fig. 1.19 shows such a load connected to the same supply as for Example 1.7.

There are several techniques which may be used to evaluate the line currents.

(1) Any unbalanced star load can be converted to an equivalent unbalanced delta load using a star/delta tranformation (see Appendix A). Thereafter proceed as for Example 1.7.

(2) Use mesh current analysis round the $\bar{I}_\alpha$ and $\bar{I}_\beta$ loops shown in Fig. 1.19.

$$\bar{E}_{RB} = \bar{I}_\alpha(\bar{Z}_a + \bar{Z}_c) - \bar{I}_\beta \bar{Z}_c$$

$$\bar{E}_{BY} = \bar{I}_\beta(\bar{Z}_c + \bar{Z}_b) - \bar{I}_\alpha \bar{Z}_c$$

Again the main problem here is to insert the correct phase relationship between $\bar{E}_{RB}$ and $\bar{E}_{BY}$.

(3) Nodal analysis taking point 0 in Fig. 1.19 as reference node. Summing currents at node S gives

$$\frac{\bar{V}_s - \bar{V}_a}{\bar{Z}_a} + \frac{\bar{V}_s - \bar{V}_b}{\bar{Z}_b} + \frac{\bar{V}_s - \bar{V}_c}{\bar{Z}_c} = 0$$

$$\therefore \quad \bar{V}_s\left(\frac{1}{\bar{Z}_a} + \frac{1}{\bar{Z}_b} + \frac{1}{\bar{Z}_c}\right) = \frac{\bar{V}_a}{\bar{Z}_a} + \frac{\bar{V}_b}{\bar{Z}_b} + \frac{\bar{V}_c}{\bar{Z}_c}$$

or

$$\text{or} \quad \bar{V}_s(\bar{Y}_a + \bar{Y}_b + \bar{Y}_c) = \bar{V}_a\bar{Y}_a + \bar{V}_b\bar{Y}_b + \bar{V}_c\bar{Y}_b \tag{1.18}$$

$\bar{V}_s$ the potential of the load star point can be found from equation (1.18) and any line current can then be found in terms of the voltage difference across the corresponding load impedance.

In cases where the load star point voltage is required then method (3) is best but where line currents only are required then there is not much to choose between methods (2) and (3). The very concise form of equation (1.18) probably gives method (3) a slight edge.

Example 1.8

For the system shown in Fig. 1.19 the supply is 400 V, 50 Hz, phase sequence RYB and $\bar{Z}_a = 5 \angle 30°\,\Omega$, $\bar{Z}_b = 5 \angle 45°\,\Omega$ and $\bar{Z}_c = 5 \angle -30°\,\Omega$. Calculate the potential of the load star point and the line currents.

The star point potential comes directly from equation (1.18) noting that

$$\bar{V}_a = \bar{E}_R = \frac{400}{\sqrt{3}} \angle 0° \text{ V (reference phasor)}$$

$$\bar{V}_b = \bar{E}_Y = \frac{400}{\sqrt{3}} \angle -120° \text{ V}$$

$$\bar{V}_c = \bar{E}_B = \frac{400}{\sqrt{3}} \angle -240° \text{ V}$$

$$\therefore \quad \bar{V}_s = \frac{400}{\sqrt{3}} \frac{(0.2 \angle -30° + 1 \angle -120° \times 0.2 \angle -45° + 1 \angle -240° \times 0.2 \angle 30°)}{0.2 \angle -30° + 0.2 \angle -45° + 0.2 \angle 30°}$$

$$= \frac{400}{\sqrt{3}}\left[\frac{0.2 \angle -165°}{0.508 \angle -16.2°}\right] = \frac{400}{\sqrt{3}} \times 0.394 \angle -148.8°$$

$$= \underline{91 \angle -148.8° \text{ V}}$$

$\bar{I}_R$ can now be evaluated from the potential difference across $\bar{Z}_a$, i.e.

$$\bar{E}_R - \bar{V}_s = \bar{I}_R\,\bar{Z}_a,\ \bar{I}_R = \frac{400}{\sqrt{3}}\ \frac{(1\ \angle\ 0^\circ - 0.394\ \angle\ -148.8^\circ)}{5\ \angle\ 30^\circ}$$

$$= \underline{62.5\ \angle\ -21.31^\circ\ \mathrm{V}}$$

$$\bar{I}_Y = \frac{\bar{E}_Y - \bar{V}_s}{\bar{Z}_b} = \frac{400}{\sqrt{3}}\ \frac{(1\ \angle\ -120^\circ - 0.394\ \angle\ -148.8^\circ)}{5\ \angle\ 45^\circ}$$

$$= \underline{31.49\ \angle\ 211.2^\circ\ \text{amperes}}$$

$$\bar{I}_B = \frac{\bar{E}_B - \bar{V}_s}{\bar{Z}_c} = \frac{400}{\sqrt{3}}\ \frac{(1\ \angle\ -240^\circ - 0.394\ \angle\ -148.8^\circ)}{5\ \angle\ -30^\circ}$$

$$= \underline{49.99\ \angle\ 128.7^\circ\ \text{amperes}}$$

The phase sequence RYB determines the angles of $\bar{E}_Y$ and $\bar{E}_B$ with respect to $\bar{E}_R$. Would the answers in Example 1.8 be different if the phase sequence was RBY ? The answer is yes since $\bar{E}_B$ is now $400/\sqrt{3}\ \angle\ -120^\circ$ instead of $400/\sqrt{3}\ \angle\ -240^\circ$ and $\bar{E}_Y$ is now $400/\sqrt{3}\ \angle\ -240^\circ$.

It is dangerous to reverse the direction of rotation of the supply phasors in order to achieve the change in phase sequence. It is much better to interchange the position of the Y and B phasors and maintain the anticlockwise direction of rotation. The reason for this caution lies in the significance of the j operator. It will always rotate a phasor 90° in an anticlockwise direction irrespective of which way the phasor diagram is rotating.

1.8.5 Measurement of power in three-phase circuits

The wattmeters W in Fig. 1.20 read the time average of the product vi where i is the current through their current coil and v is the potential difference across their voltage coil. The current coil is shown as a loop in series with the line and the voltage coil is shown as a resistor across the potential difference in question. We shall assume that the potential difference across the current coil and the current drawn by the potential coil are both negligible.

The power drawn by the star-connected load in Fig. 1.20 when the currents and voltages are periodic functions of ω is

$$P = \frac{1}{2\pi}\int_0^{2\pi} (v_{oa}\cdot i_a + v_{ob}\cdot e_b + v_{oc}\cdot e_c)\ \mathrm{d}(\omega t) \qquad (1.19)$$

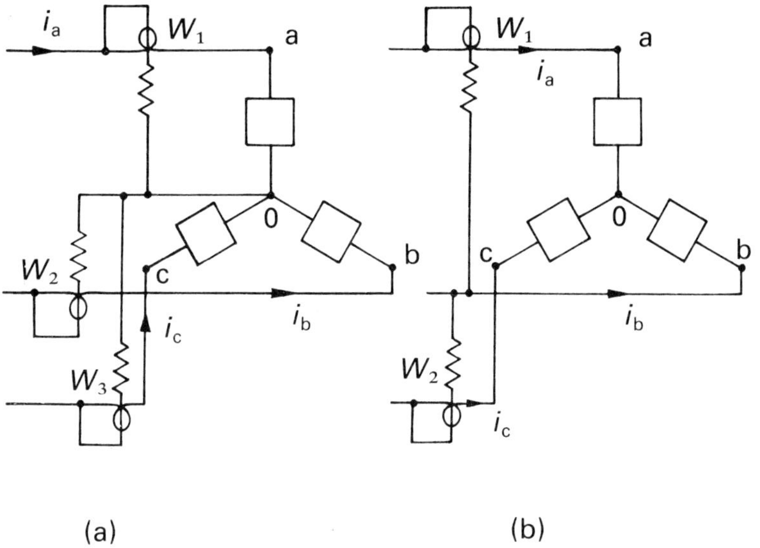

Fig. 1.20 Measurement of three-phase power using (a) three wattmeters and (b) two wattmeters.

The three wattmeter connection shown in Fig. 1.20(a) performs this task directly with each wattmeter reading one of the terms on the right-hand side of equation (1.19).

When the load is supplied via three wires the total power can be read on two wattmeters connected as in Fig. 1.20(b).

$$W_1 = \frac{1}{2\pi}\int_0^{2\pi} (v_{ba} \cdot i_a)\, d(\omega t), \qquad W_2 = \frac{1}{2\pi}\int_0^{2\pi} (v_{bc} \cdot i_c)\, d(\omega t)$$

and

$$W_1 + W_2 = \frac{1}{2\pi}\int_0^{2\pi} \left[v_{ba} \cdot i_a + v_{bc} \cdot i_c \right] d(\omega t)$$

Now express v_{ba} and v_{bc} in terms of phase voltages to get

$$W_1 + W_2 = \frac{1}{2\pi}\int_0^{2\pi} \left[(v_{bo} + v_{oa}) \cdot i_a + (v_{bo} + v_{oc}) \cdot i_c \right] d(\omega t)$$

$$= \frac{1}{2\pi}\int_0^{2\pi} \left[v_{oa} \cdot i_a + v_{oc} \cdot i_c + v_{bo} \cdot (i_a + i_c) \right] d(\omega t)$$

But at node 0,

$$i_a + i_b + i_c = 0$$

and therefore

$$(i_a + i_c) = -\, i_b.$$

With

$$v_{ob} = -v_{bo}$$

we now have

$$W_1 + W_2 = \frac{1}{2\pi}\int_0^{2\pi} (v_{oa} \cdot i_a + v_{ob} \cdot i_b + v_{oc} \cdot i_c)\, d\omega t.$$

(Q.E.D.)

The proof has been carried out for a star load but is equally true for a delta load or indeed any load supplied via three wires. There is no need for the load to be balanced or for the variables to be sinewaves but they must be harmonic functions of ω.

When the two wattmeter method is applied to a balanced three-phase load as in Fig. 1.21 certain relationships exist between the readings W_1 and W_2.

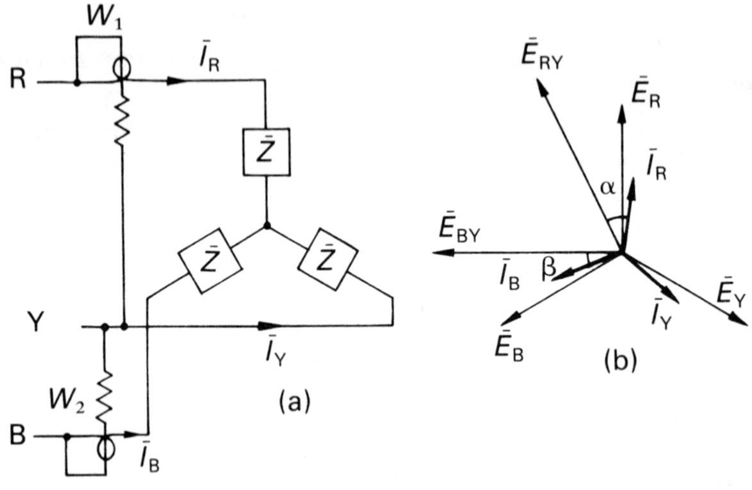

Fig. 1.21 Wattmeter connection (a) and phasor diagram (b) for a balanced three-phase star connected load.

From Section 1.4.2 W_1 measures the real part of $\bar{E}_{RY}.\bar{I}_R^*$. Hence W_1 reads $E_{RY}.I_R \cos\alpha = E_{RY}.I_R \cos(30^\circ + \theta)$, where θ is the load phase angle.

Similarly, W_2 reads

$$E_{BY} \cdot I_B \cos\beta = E_{BY} \cdot I_B \cos(30^\circ - \theta)$$

$$\therefore \quad W_1 + W_2 = V_L I_L (\cos(30^\circ + \theta) + \cos(30 - \theta))$$

$$= \underline{\sqrt{3}\; V_L I_L \cos\theta}$$

also $$W_1 - W_2 = V_L I_L \left[\cos(30^\circ + \theta) - \cos(30^\circ - \theta) \right]$$

$$= - V_L I_L \sin\theta$$

$$\frac{W_1 - W_2}{W_1 + W_2} = - \frac{1}{\sqrt{3}} \tan\theta$$

(Note that positive θ indicates lagging current.)

Hence in the balanced load case the load phase angle can be deduced from the two wattmeter readings, provided we know which of the two readings is designated as W_1. The convention we have chosen is that W_1 is the wattmeter involving a particular line current and the potential difference between that line and the *following line in the phase sequence.*

Example 1.9
For the system of Fig. 1.18 two wattmeters are connected as shown in Fig. 1.22.

Calculate the wattmeter readings.

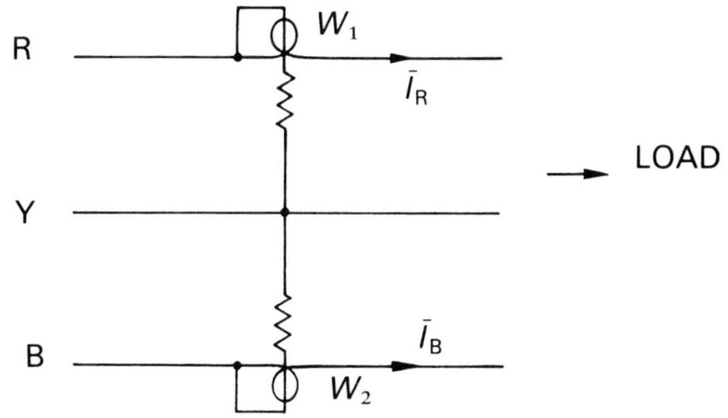

Fig. 1.22 Wattmeter connections for the unbalanced delta load shown in Fig. 1.18.

$$W_1 = E_{RY} \cdot I_R \cos\alpha$$

where α is the angle between $\bar{E}_{RY}$ and $\bar{I}_R$.

From the solution to Example 1.7 $\alpha = 14.5^\circ$, since $\bar{E}_{RY}$ is the reference phasor. Hence

$$W_1 = 400 \times 7.5 \cos 14.5^\circ = \underline{2904.4 \text{ W}}$$

$$W_2 = E_{BY} \cdot I_B \cdot \cos\beta$$

where β is the angle between $\overline{E}_{BY}$ and $\overline{I}_B$.

$$\overline{E}_{BY} = V_L \angle 60^\circ \text{ V and } \overline{I}_B = 4.3 \angle -47.2^\circ \text{ amperes}$$

Hence,

$$\beta = 107.2^\circ$$

and

$$W_2 = 400 \times 4.3 \times \cos 107.2$$

$$= \underline{-509 \text{ W}}$$

Hence the total power $W_1 + W_2 = \underline{2395 \text{ W}}$

Compare this with the 2399 W calculated in Example 1.7. The difference is due to rounding errors in the currents. Note that W_2 is negative and that the total power is the algebraic sum of W_1 and W_2.

PROBLEMS

1 Three parallel branches carry the following currents

$$i_1 = 10 \sin 314t \text{ amperes,} \qquad i_2 = 7 \sin (314t - \frac{\pi}{3}) \text{ amperes,}$$

$$i_3 = 12 \sin (314t + \frac{\pi}{4}) \text{ amperes.}$$

Sketch a phasor diagram and find an expression for the total current. What is the r.m.s. value of the total current and what is its frequency?

Ans. $22.14 \sin (314t + 0.11)$ amperes, 15.65 amperes, 50 Hz.

2 An e.m.f. given by $326 \sin 418t$ volts is applied to a resistance and inductance in series. The current drawn is $2 \sin (418t - 1.3736)$ A. What is the supply frequency and the resistance and inductance of the circuit? Calculate the r.m.s. values of the voltage and current and the average power drawn by the circuit. (The current phase angle is given in radians.)

Ans. 66.53 Hz, 31.94 Ω, 0.382 H, 230.5 V, 1.414 A, 63.9 W.

3 A projector lamp rated at 1000 W, 115 V is to be run from a 240 V, 50 Hz supply, by connecting it in series with an inductor. If the ratio of the inductor resistance to its reactance is 0.1 find the resistance and reactance of the inductor to run the lamp at its correct rating. (The lamp is a pure resistance.) What capacitor connected across the supply will make the supply power factor unity?

Ans. 2.28 Ω, 22.8 Ω, 95.5 μF.

4 A delta connected load is made up of the following impedances. $\bar{Z}_{RY} = 65 \angle 30^\circ\ \Omega$, $\bar{Z}_{YB} = 80 \angle 50^\circ\ \Omega$, $\bar{Z}_{BR} = 40 \angle -20^\circ\ \Omega$. It is connected to a 400 V, 50 Hz, three-phase supply of phase sequence RYB. Calculate the line currents.

Ans. $\bar{I}_R = 16.1 \angle 24^\circ$ A, $\bar{I}_Y = 10.5 \angle 228^\circ$ A, $\bar{I}_B = 7.4 \angle 170^\circ$ A ($\bar{E}_{RB}$ reference phasor).

5 A star connected load is made up of the following impedances. $\bar{Z}_R = 45 \angle 0^\circ\ \Omega$, $\bar{Z}_Y = 60 \angle 60^\circ\ \Omega$, $\bar{Z}_B = 55 \angle 30^\circ\ \Omega$. It is connected to a 600 V, 50 Hz, three-phase supply of phase sequence RYB. Calculate the line currents.

Ans. $\bar{I}_R = 8.64 \angle 11.29$ A, $\bar{I}_Y = 7.51 \angle -139.61$ A, $\bar{I}_B = -(\bar{I}_R + \bar{I}_Y) = 4.21 \angle 130.82$ A ($\bar{E}_{RB}$ reference phasor).

6 A balanced star load consists of three impedances of $\bar{Z}\,\Omega$. Show that the equivalent delta load has impedances of $3\bar{Z}\,\Omega$.

A 415 V, 50 Hz, three-phase supply is connected to a balanced star load of impedance per phase $(8 + j6)\,\Omega$ by an interconnector of impedance per phase $(0.1 + j0)\,\Omega$. Calculate the power dissipated in the interconnector. The load power factor is brought to unity by connecting capacitors of magnitude C in delta across the load. Calculate C and the loss in the interconnector with the capacitors in circuit.

Ans. 169.5 W, 16.6 μF, 110 W.

7 Calculate the readings on two wattmeters connected to read the total power in the circuit of Problem 5.

Ans. 5.084 kW, 813 W.

8 Two wattmeters are used to measure the power in a balanced 440 V, 50 Hz, three-phase system. The indication of W_2 and W_1 are 42.6 kW and 31.2 kW respectively. Calculate the line current and the power factor of the load.

Ans. 100.5 A, 0.966 lagging.

2

Electromagnetism for machines

2.1 INTRODUCTION

This chapter will deal with the application of simple field theory to situations relevant to the study of electrical machines and transformers in the steady state. It will be assumed that the reader is familiar with elementary electromagnetic field theory which covers the following prerequisities.

(1) Faraday's law: $\text{e.m.f.} = -N\dfrac{\mathrm{d}\phi}{\mathrm{d}t}$;

$$\text{e.m.f.} = -L\frac{\mathrm{d}i_1}{\mathrm{d}t} \pm M\frac{\mathrm{d}i_2}{\mathrm{d}t};$$

$$\text{e.m.f.} = Blv$$

when B is not a function of time and B, l and v are mutually perpendicular.

(2) Ampère's circuital law $\oint \mathbf{H}\cdot\mathbf{d}l = IN$; $\mathbf{B} = \mu_r\mu_0\mathbf{H}$.

(3) Stored magnetic energy in a two current system is $\frac{1}{2}L_1I_1^2 + \frac{1}{2}L_2I_2^2 \pm MI_1I_2$.
Stored energy in magnetic field $= \int_v(\frac{1}{2}BH)\,\mathrm{d}V = W_m$

(4) Torque in a magnetic system is given by $T = \partial W_m/\partial\theta$, where W_m is the stored magnetic energy.

(5) Force on a conductor l carrying a current I in a magnetic field B is BlI, when B, l and I are mutually perpendicular.

It is clear from the prerequisites that there are no electrostatic machines of general significance. It is difficult to produce any equivalent force which is comparable with a BlI force without resorting to complicated and expensive apparatus.

2.2 THE MAGNETIC CIRCUIT

2.2.1 Ampère's law

In an electric circuit it requires an electromotive force (e.m.f.) to set up a current. In a magnetic circuit it requires a magnetomotive force (m.m.f.) to set up a flux.

It was Oersted who first discovered that an electric current sets up a magnetic field and this effect was later formalised into Ampère's circuital law, $\oint \mathbf{H} \cdot \mathbf{dl} = IN$ and later still into one of Maxwell's equations curl $\mathbf{H} = \mathbf{J}(+\dot{\mathbf{D}})$. Ampère's law will suffice for the simple field configurations dealt with in this text. In words, what Ampère's law tells us is that the line integral of the magnetic field intensity $\mathbf{H}$ taken round a closed contour ($\oint \mathbf{H} \cdot \mathbf{d}l$) is equal to the current enclosed by this contour, (IN). $\oint \mathbf{H} \cdot \mathbf{d}l$ is called the m.m.f. (cf. $\oint \mathbf{E} \cdot \mathbf{dl} \rightarrow$ e.m.f.) and is measured in units referred to as ampere-turns (AT). The 'turns' in ampere-turns is dimensionless and the fundamental unit in which m.m.f. is measured is therefore amperes.

Fig. 2.1 illustrates two simple applications of Ampère's law.

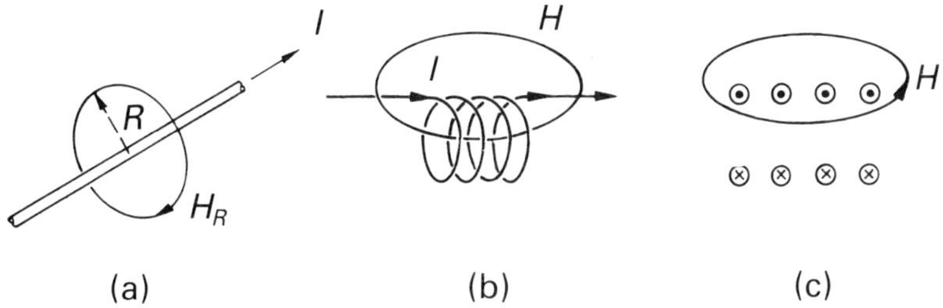

Fig. 2.1 A typical flux line for (a) an infinite wire and (b), (c) a coil.

In Fig. 2.1(a) an infinitely long wire carrying current I sets up flux lines which are concentric with the conductor. If we integrate round the flux line shown then $\mathbf{H} \cdot \mathbf{d}l$ becomes $H \cdot dl$ and since H_R is uniform along the path shown $\oint H \cdot dl = 2\pi R H_R$. The current enclosed is clearly I and hence $H_R = I/2\pi R$.

Fig. 2.1(b) shows another common situation, that of a coil carrying a current I setting up flux lines 'linking' the coil and returning outside the turns. $\oint \mathbf{H} \cdot \mathbf{d}l$ along the flux line shown will again become $\oint H \cdot dl$ but H is not uniform along the contour and the evaluation of $\oint H \cdot dl$ would require information about the space variation of H. The current enclosed is best seen by reference to Fig. 2.1(c) which shows a section through the coil in the plane of the flux line. Inside the contour we can see four conductors carrying current I and hence the amperes enclosed is $4I$. This explains why we use ampere-turns as the unit of m.m.f. since the most common 'source' of flux is an N turn coil.

The examples of Fig. 2.1 serve to illustrate that the calculation of H from Ampère's law is only straightforward in situations where we already have information about the shape and density of the flux lines. Fortunately these conditions apply to most of the situations we need consider.

2.2.2 Reluctance of a flux path

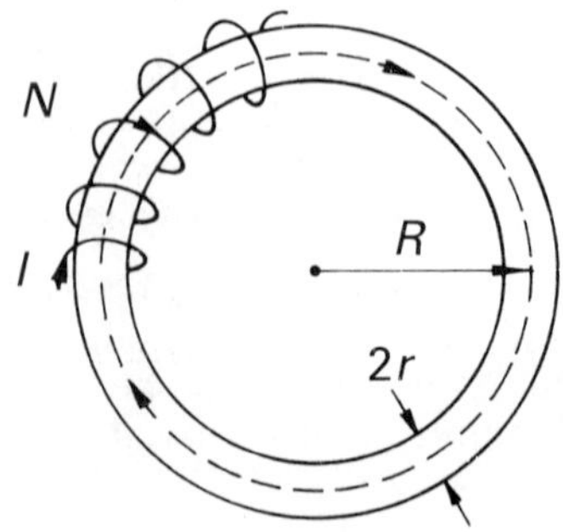

Fig. 2.2 A toroid of magnetic material carrying N uniformly spaced turns.

Fig. 2.2 shows a toroid of magnetic material with a winding of N turns uniformly distributed around its circumference. Only part of the winding is shown in Fig. 2.2 and the dotted line shows a typical flux loop at the mean toroid radius of R. Provided μ_r for the magnetic material is much greater than 1 then the lines of flux will remain inside the toroid and will form concentric circles. This is the sort of information we require about the spatial variation of **H** to allow us to use Ampère's law to evaluate **H**.

Consider a flux line at the mean radius R and integrate **H** round this line.

$$\oint \mathbf{H} \cdot \mathbf{d}l = \oint H_R \, dl = 2\pi R H_R$$

and from Ampère's law

$$2\pi R H_R = IN, \qquad H_R = \frac{\mathrm{IN}}{2\pi R} \tag{2.1}$$

The value of H will vary slightly as we consider flux loops away from the mean radius R. If the radius r of the toroid section is much less than R then to a reasonable approximation H within the toroid, at any radius, is given by equation (2.1). The direction of H is circumferential.

The flux through the toroid section is $\int \mathbf{B} \cdot \mathbf{d}A$ over the toroid section, where $\mathbf{B} = \mu_r \mu_0 \mathbf{H}$. **B** like **H** is constant over the section and hence

$$\phi = BA = \mu_r \mu_0 HA$$

Substituting for H from equation (2.1) gives

$$\phi = \mu_r \mu_0 \frac{IN}{2\pi R} \cdot A = IN \frac{(\mu_r \mu_0 A)}{2\pi R}$$

and hence

$$IN = \phi \times \left(\frac{2\pi R}{\mu_r \mu_0 A}\right) \tag{2.2}$$

Equation (2.2) is in the form of m.m.f. = (flux) × $(2\pi R/\mu_r \mu_0 A)$ which can be compared with Ohm's law for electric circuits

e.m.f. = current × resistance

The quantity in brackets in equation (2.2) is referred to as the 'reluctance' of the magnetic circuit and is given the symbol $\mathcal{R}$. Equation (2.2) now becomes

$$IN = \phi \times \mathcal{R} \tag{2.3}$$

In a magnetic circuit when the flux density is uniform then $\mathcal{R}$ is always of the form

$$\left(\frac{\text{length of magnetic circuit}}{\mu_r \mu_0 \times \text{cross-section area}}\right)$$

which is analogous with $R = l/\sigma A$ for resistance. Not surprisingly the laws for reluctances in series and parallel are exactly the same as for resistance in series and parallel.

Example 2.1

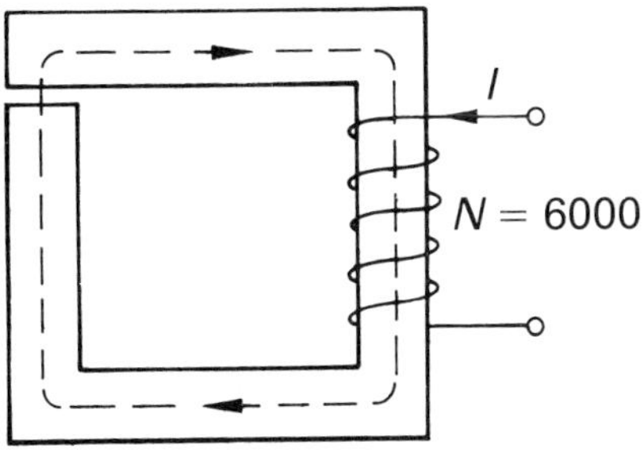

Fig. 2.3 A magnetic circuit with a small air-gap.

Fig. 2.3 shows an iron circuit with a small air-gap cut in it. A 6000 turn coil carries a current $I = 20\,\text{mA}$ which sets up a flux within the iron and across the air gap. If the iron cross-section is $0.8 \times 10^{-4}\,\text{m}^2$, the mean length of flux path in iron is 0.15 m, $\mu_r = 800$ in iron and air-gap length is 0.75 mm, calculate the air-gap flux and the energy stored in the air-gap magnetic field. It may be assumed that the flux lines flow straight across the air-gap, i.e. air-gap cross-section is also 0.8×10^{-4} m^2.

Here we have two reluctances in series carrying the same flux. The total reluctance $= \mathcal{R}_1 + \mathcal{R}_2$.

$$\text{For the iron } \mathcal{R}_1 = \frac{l}{\mu_r \mu_0 A} = \frac{0.15}{800 \times 4\pi \times 10^{-7} \times 0.8 \times 10^{-4}}$$

$$= \underline{18.65 \times 10^5 \text{ AT/Wb}}$$

$$\mathcal{R}_2 \text{ (air-gap)} = \frac{0.75 \times 10^{-3}}{1 \times 4\pi \times 10^{-7} \times 0.8 \times 10^{-4}}$$

$$= 0.75 \times 10^7 = \underline{75 \times 10^5 \text{AT/Wb}}$$

$$\text{Hence } \mathcal{R}_1 + \mathcal{R}_2 = \underline{93.75 \times 10^5 \text{AT/Wb}}$$

$$\text{The flux } \phi = \frac{\text{m.m.f.}}{\mathcal{R}_1 + \mathcal{R}_2} = \frac{6000 \times 20 \times 10^{-3}}{93.75 \times 10^5}$$

$$= \underline{12.8 \times 10^{-6} \text{ Webers}}$$

$$\text{The air-gap flux density} = \frac{\phi}{0.8 \times 10^{-4}} = \underline{0.16 \text{ Tesla}}$$

B, and therefore H, is uniform in the air-gap and hence the stored energy in the air-gap $= (\frac{1}{2}BH) \times$ air-gap volume

$$\therefore \quad \text{Stored energy} = \tfrac{1}{2} \times 0.16 \times \frac{0.16}{4\pi \times 10^{-7}} \times 0.8 \times 10^{-4} \times 0.75 \times 10^{-3}$$

$$= \underline{0.611 \times 10^{-3} \text{ joules}}$$

Example 2.1 makes the assumption that all the flux linking the coil crosses

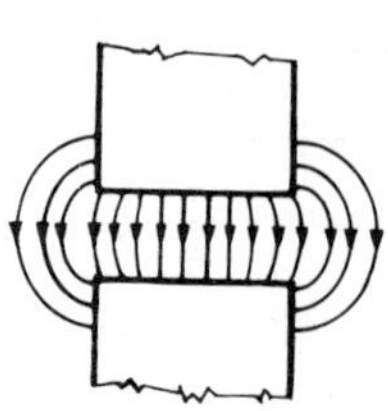

Fig. 2.4 Fringing of the magnetic field around the air-gap.

the air-gap, i.e. there is no flux 'leakage' and further that there is no 'fringing' of the magnetic field at the air-gap. Fig. 2.4 shows what is meant by fringing and it has the effect of reducing the flux density in the air-gap.

2.3 THE *B*/*H* RELATIONSHIP IN IRON

A proper treatment of the magnetic properties of iron is beyond the scope of this text. What information we require for the solution of magnetic circuit problems is contained in the simple *B*/*H* relationship given in Fig. 2.5.

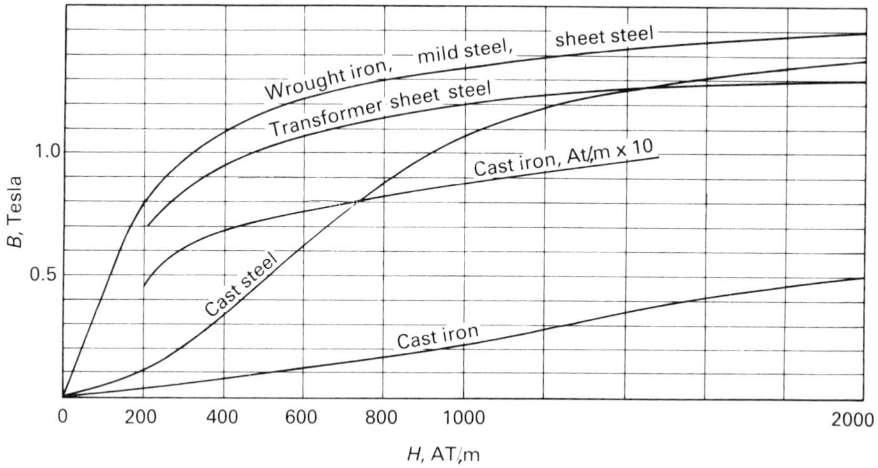

Fig. 2.5 *B*/*H* curves for common magnetic materials.

Hitherto we have assumed a linear relationship between *B* and *H* given by $B = \mu_r\mu_0 H$. $\mu_r\mu_0$ is the slope of *B*/*H* characteristics given in Fig. 2.5 and it is clear that above a flux density of approximately 0.8 T for, say, transformer sheet steel, μ_r is not constant. This will lead to a non-linear flux/reluctance relationship and thereby complicate the solution of magnetic circuit problems. Provided the flux density is such that the *B*/*H* relationship is in the linear region then the solution of a magnetic circuit is exactly as for a linear electric circuit. When the flux density reaches a level at which the *B*/*H* relationship is non-linear then the problem becomes analogous to the electrical circuit with a non-linear resistance.

Example 2.2

If the material in the iron circuit of Example 2.1 is mild steel what is the air-gap flux density for a current of (1) 20 mA, (2) 120 mA?

Let us assume that for 20 mA the flux density in the mild steel is less than

0.8 T and that we can make use of the reluctance concept. The slope of the B/H curve in the linear region is $1/220 = \mu_r\mu_0$.

$$\text{Hence } \mathcal{R}_1 \text{ (iron)} = \frac{0.15 \times 220}{0.8 \times 10^{-4}} = 4.125 \times 10^5 \text{ AT/Wb}$$

$$\mathcal{R}_2 \text{ as before} = 75 \times 10^5 \text{ AT/Wb}$$

$$\therefore \quad \mathcal{R}_1 + \mathcal{R}_2 = 79 \times 125 \times 10^5 \text{ AT/Wb}$$

$$\therefore \quad \phi = \frac{6000 \times 20 \times 10^{-3}}{79.125 \times 10^5} = 1.517 \times 10^{-5} \text{ Wb}$$

and

$$B = \frac{\phi}{A} = \frac{1.517 \times 10^{-5}}{0.8 \times 10^{-4}} = \underline{0.19 \text{ Tesla}}$$

This answer validates our initial assumption that B is in the linear region of the B/H curve.

In the second case I and hence the m.m.f. increase by a factor of 6. If both $\mathcal{R}_1$ and $\mathcal{R}_2$ were linear functions of B then B would also increase six-fold to 1.14 T. $\mathcal{R}_2$ is linear with B and it is clear from part 1 that $\mathcal{R}_2$ dominates the total reluctance value. A slight departure from linearity for $\mathcal{R}_1$ will not greatly affect the answer for B but it will substantially affect H_i as we shall see.

Returning to Ampère's law and integrating round the dotted contour in Fig. 2.3 we have

$$\oint \mathbf{H} \cdot \mathbf{d}l = \underset{\text{(iron)}}{H_i l_i} + \underset{\text{(air)}}{H_a l_a} = 6000 \times 0.12 = 720 \text{ AT}$$

but

$$H_a = \frac{B_a}{\mu_0} \quad \text{and} \quad B_i = B_a = B$$

$$\therefore \quad H_i \times 0.15 + \frac{B}{\mu_0} \times 0.75 \times 10^{-3} = 720$$

$$\therefore \quad B = 1.206 - 2.51 \times 10^{-4} H_i \qquad (1)$$

Equation (1) gives a straight line relationship which can be plotted on Fig. 2.5. Where this line cuts the B/H curve for mild steel gives the solution for B. The reader should verify that the intersection takes place at $B \simeq 1.09$ Tesla giving $H_i = 400$ AT/m. This value for B is only 5 per cent less than for the linear case but the value for H_i in the linear case would have been about 220 AT/m. (Continue initial slope of B/H curve.)

* * *

Provided the air-gap flux density remains in the region up to the 'shoulder' of the B/H curve the linear model is a reasonable approximation for the calculation of B in situations where the air-gap reluctance is much greater than the reluctance of the iron path. These conditions will apply for the simple machine types covered in this text.

2.4 FLUX-DENSITY DISTRIBUTION IN A CYLINDRICAL AIR-GAP MACHINE

The magnetic circuit of a cylindrical air-gap radial flux machine is essentially that presented by two iron cylinders mounted concentrically one inside the other as in Fig. 2.6(a).

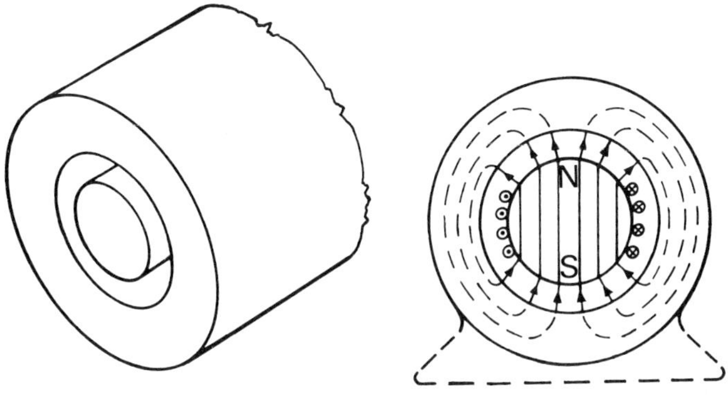

(a) (b)

Fig. 2.6 Basic structure of the uniform air-gap radial flux machine. (a) Concentric iron cylinders and (b) radial flux lines set up by axial currents as shown.

The air-gap flux is produced by current carrying conductors running parallel to the cylinder axis and placed on the surface of one or other cylinder. Fig. 2.6(b) shows a section at right angles to the cylinder axis with a band of current carrying conductors placed symmetrically on either side of the inner cylinder. The current is flowing into the page on the RHS and returning round the far end of the inner cylinder to flow up out of the page on the LHS. The flux lines cross the air-gap radially and form a symmetrical pattern as shown. The outer cylinder, called the stator, is usually supported by a base structure as shown dotted in Fig. 2.6(b) while the inner cylinder, the rotor, runs on a shaft supported by bearings fixed to the outer cylinder at both ends. The air-gap between the cylinders is shown much exaggerated in Fig. 2.6(b). A typical ratio of air-gap radial length to inner cylinder radius would be of the order 1/50. Given the flux pattern of Fig. 2.6(b) it is possible to calculate H and hence B in the air-gap. Because the air-gap is much exaggerated in Fig. 2.6(b) the flux lines diverge appreciably. In reality the divergence is much smaller and we shall consider H and B to be constant in the radial direction in the air-gap. One further geometrical manoeuvre which we shall use extensively is to 'unroll' or develop the air-gap into a flat geometry as shown in Fig. 2.7(a). Again because of the small

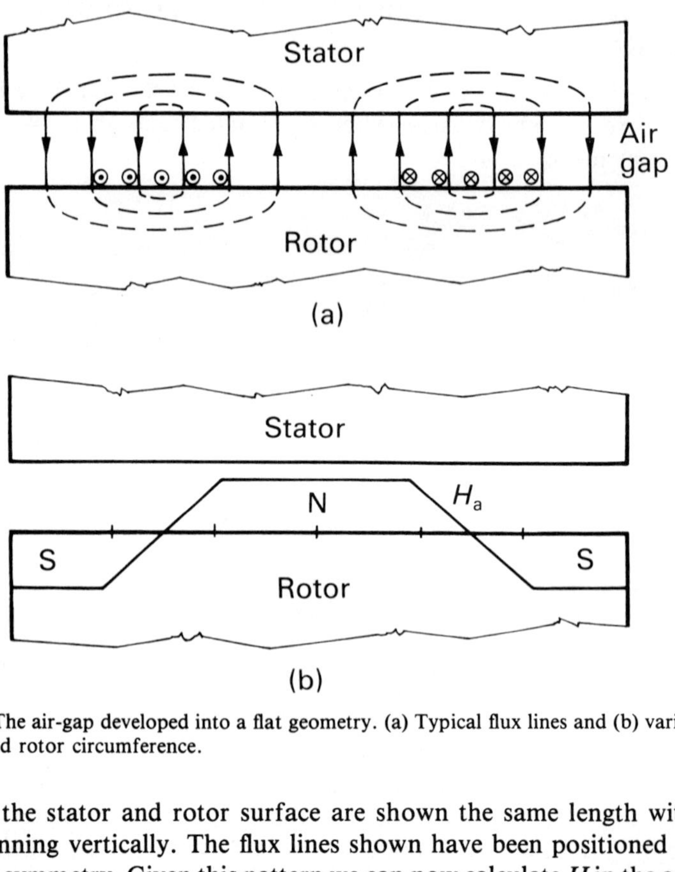

Fig. 2.7 The air-gap developed into a flat geometry. (a) Typical flux lines and (b) variation of H_a around rotor circumference.

air-gap the stator and rotor surface are shown the same length with flux lines running vertically. The flux lines shown have been positioned on the basis of symmetry. Given this pattern we can now calculate H in the air-gap.

Consider the outer loop on the LHS of Fig. 2.7(a). Ampère's law round this loop gives

$$H_i \cdot l_i + H_a \cdot l_a = IN$$

where IN is the enclosed current.

Now as we found in Examples 2.2 and 2.3 $H_i l_i \ll H_a l_a$ and to a reasonable approximation we can say

$$H_a l_a = IN$$

l_a in this case is $2g$, where g is the radial length of the air-gap.

Hence $H_a \doteq IN/2g$ for any loop which encloses all the turns.

For the inner loop which does not enclose all the current H_a must clearly decrease. If the ampere turn distribution is linear within the current carrying region then the value of H_a will be proportional to the width of the loop. H_a will fall linearly to zero at the centre of the current carrying region and will be equal to $\pm(IN/2g)$ at the ends of the region. Fig. 2.7(b) shows the resulting distribution of H_a along the air-gap taking downward pointing flux as negative. $B_a = \mu_0 H_a$ will have the same distribution in the air-gap as shown in Fig. 2.7(b). This type of flux distribution would be referred to as a 2-pole distribution having one maximum (north) and one minimum (south) region. Fig. 2.6(b) is perhaps the better diagram to observe in order to see the 2-pole arrangement.

If there is an axial conductor of length l on the inner surface of the stator then as the rotor rotates at ω rad/s the e.m.f. generated in the conductor can be calculated from the $B_a l v$ rule. B_a is not a function of time if the rotor winding current is constant. Since l and v are constant the variation of

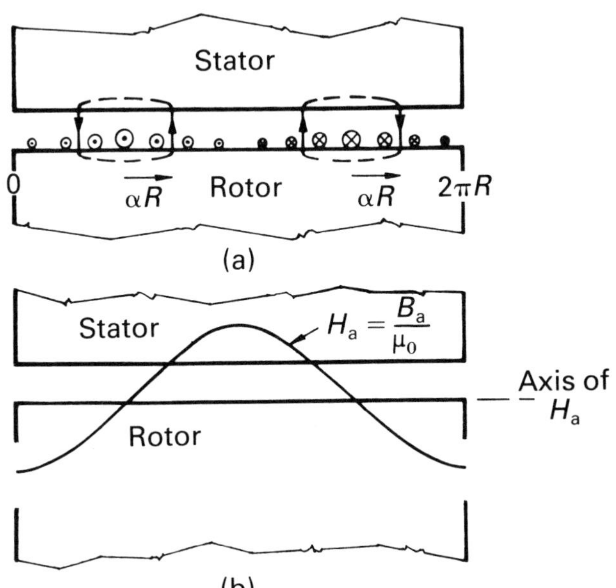

Fig. 2.8 (a) A sinusoidal surface current distribution on the rotor surface and (b) the resulting flux density distribution in the air-gap.

e.m.f. with time has the same shape as the air-gap flux density distribution. The e.m.f. will be periodic but not sinusoidal.

It is desirable to have a sinusoidal e.m.f. waveform and from the $B_a lv$ argument above it follows that this will require a sinusoidal B_a distribution in the air-gap. (See Chapter 4.) A sinusoidal B_a distribution in a uniform air-gap machine is set up by a cosinusoidal ampere-turn distribution on the rotor. Fig. 2.8(a) illustrates such a current distribution by showing conductors whose envelope conforms to a sine wave.

The flux loops will again be symmetrically distributed around the current maxima and two such loops are shown. As α increases for the given flux loops the loops will enclose more current until $\alpha = \pi/2$. Maximum H_a will therefore occur at $\alpha = \pi/2$. Zero H_a occurs when $\alpha = 0$, i.e. at the current maxima (minima). In between these two values for α the loop encloses a current of the form

$$\int_{-\alpha R}^{+\alpha R} \hat{J} \cos \alpha R \, \mathrm{d}(\alpha R),$$

where $\hat{J} \cos \alpha R$ describes the current distribution with $\alpha = 0$ corresponding to the current maxima. This gives

$$2gH_a = -2\hat{J} \sin \alpha R, \qquad H_a = \left[\frac{-\hat{J} \sin \alpha R}{g}\right]$$

The direction of H_a is given by the right-hand corkscrew rule and with the convention that downward H_a is negative Fig. 2.8(b) shows the H_a and B_a distribution in the air-gap.

There is a 90° displacement between the current distribution of Fig. 2.8(a) and the flux distribution of Fig. 2.8(b).

Note once more that the solution for H_a depended on a knowledge of the direction and symmetry of the field distribution.

2.5 TORQUE BETWEEN TWO SINUSOIDAL AMPERE-TURN DISTRIBUTIONS

Consider the case where both rotor and stator carry sinusoidal ampere-turn distributions. The rotor current and flux distributions are as for Fig. 2.8(b) and the stator current distribution is as in Fig. 2.9.

There is an angle of β between the corresponding points on the two current distributions. If the left-hand end of Fig. 2.9 is taken as reference then the stator current distribution is $\hat{J}_s \sin(\alpha - \beta)$, the rotor current distribution is $\hat{J}_r \sin \alpha$ and the rotor flux density distribution is $-\hat{B}_r \cos \alpha$.

$\hat{J}_s$ and $\hat{J}_r$ are respectively the peak surface current density in ampere-turns per unit length for the stator and rotor current distributions. R as before

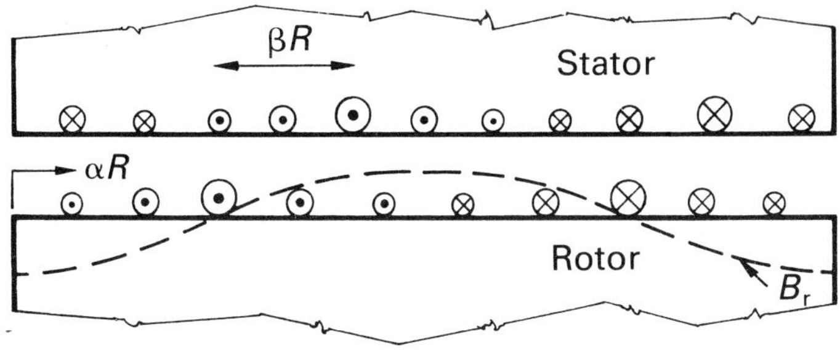

Fig. 2.9 Sinusoidal surface current distribution on both rotor and stator.

is the rotor radius and since the air-gap is of negligible length w.r.t. R the stator inner radius is also approximated to R.

The force on the stator current distribution can be calculated by use of the $B_r l I_s$ formula and Fleming's left-hand rule (direction). l is the axial length of the stator conductor (into page).

Torque on an element d (αR) of stator current distribution is

$$\delta T = R \times \underset{(I_s)}{\hat{J}_s \sin(\alpha-\beta)\ \mathrm{d}\ (\alpha R)} \cdot \underset{(B_r)}{(-\hat{B}_r)\cos\alpha} \cdot \underset{(l)}{l}$$

$$\therefore \quad \text{total torque } T = -R^2 l \hat{B}_r \hat{J}_s \int_0^{2\pi} \sin(\alpha-\beta)\cos\alpha \cdot \mathrm{d}\alpha$$

$$= \hat{B}_r \hat{J}_s l \cdot \pi R^2 \sin\beta \tag{2.4}$$

The expression in (2.4) looks as if it has an extra R in it but remember that the units of $\hat{J}_s$ are ampere-turns per unit length.

The direction of the torque from the left-hand rule, taking upward B as positive and I out of the paper as positive, is to the left. The stator of course cannot move under the action of this torque. The rotor will experience an equal and opposite torque and will therefore move to the right if free to do so.

βR in Fig. 2.9 is the distance between the two current distribution peaks. In the actual geometry of the machine as in Fig. 2.6(b) β is the corresponding angle between the current distribution peaks and the total torque T can be expressed as

$$T = K\hat{B}_r \hat{J}_s \sin\beta \tag{2.5}$$

where K is a constant of proportionality. Since B_r and B_s are each shifted 90° from their respective current distribution β is also the angle between the peaks of the two flux distributions B_r and B_s. T can also be expressed as

$$T = K'\,\hat{B}_r \hat{B}_s \sin\beta \tag{2.6}$$

since we saw earlier in Fig. 2.8 that $\hat{B}_s$ is proportional to $\hat{J}_s$. Equation (2.6) will be used frequently in later chapters to express torque in terms of the two sinusoidal flux density distributions.

2.6 TORQUE DERIVED FROM ENERGY CONSIDERATION

The previous section dealt with torque in terms of sinusoidal flux and current distributions. Sinusoidal ampere-turn distributions are impracticable and Chapter 4 shows some of the approximate methods used to derive 'near sinusoidal' ampere-turn distributions. Such distributions are periodic and consist of a fundamental and several space harmonics. The same sort of approximation is true for flux density distributions.

Where both flux and current distributions are periodic but non-sinusoidal, equations (2.5) and (2.6) can be applied to terms of like frequency in the Fourier series for the flux and current distributions. The calculation of torque by such methods requires a detailed knowledge of the machine design and uses variables not measurable at the machine terminals. This latter point is very important to the system engineer who wishes to be able to describe the machine in a system by an equivalent circuit.

Let us now use the expression $T = \partial W_m/\partial\theta$, where W_m is the stored magnetic energy in the two coil (rotor and stator winding) system of the machine. W_m may be expressed in terms of the winding inductances and currents as

$$W_m = \tfrac{1}{2}L_s I_s^2 + \tfrac{1}{2}L_r I_r^2 + M I_s I_r \tag{2.7}$$

I_s and I_r are independent of θ but L_s, L_r and M may be functions of θ.

$$\frac{\partial W_m}{\partial\theta} = \tfrac{1}{2}I_s^2\frac{\partial L_s}{\partial\theta} + \tfrac{1}{2}I_r^2\frac{\partial L_r}{\partial\theta} + I_s I_r\frac{\partial M}{\partial\theta} \tag{2.8}$$

In the uniform air-gap machine the self-inductance of any winding does not change with θ. The shape of the flux path and hence its reluctance do not change with rotor position. Equation (2.5) then reduces to

$$\frac{\partial W_m}{\partial\theta} = I_s I_r\frac{\partial M}{\partial\theta} \tag{2.9}$$

The torque in a given machine must be the same irrespective of what analytical technique we use to evaluate it. Equation (2.9) must therefore be equivalent to the *BlI* derived formulae of the previous section. Let us illustrate this equivalence with respect to a simple machine model as in Fig. 2.10(a).

Fig. 2.10(b) shows the development of Fig. 2.10(a). The flux density profile we know from previous analysis.

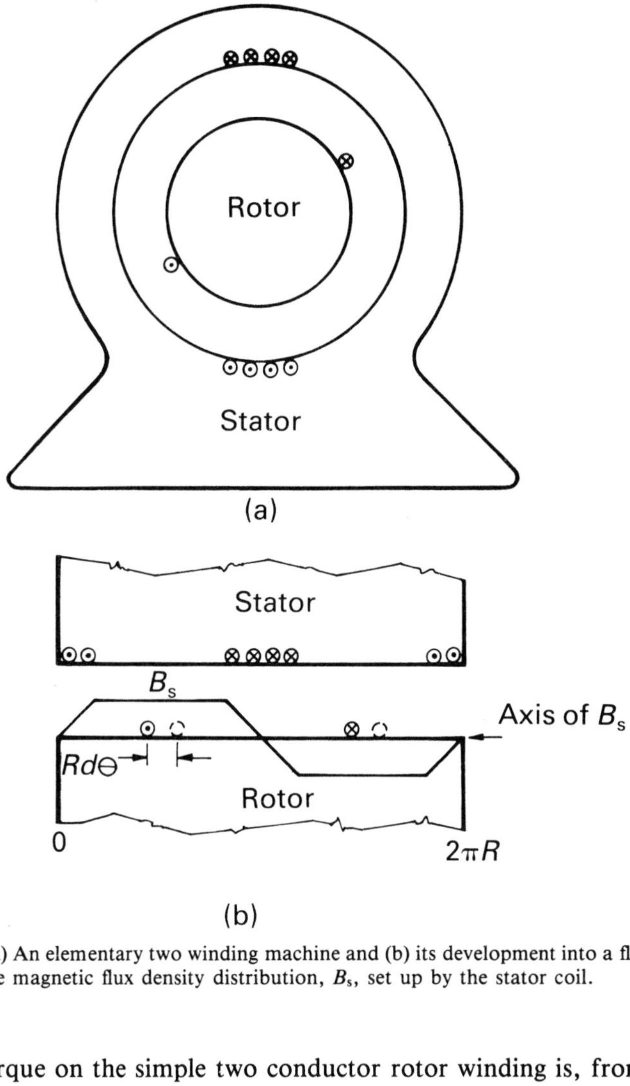

(a)

(b)

Fig. 2.10 (a) An elementary two winding machine and (b) its development into a flat geometry showing the magnetic flux density distribution, B_s, set up by the stator coil.

The torque on the simple two conductor rotor winding is, from the *BlI* rule,

$$T = 2\hat{B}_s l I_r \times R$$

anticlockwise, where the symbols have their usual meaning.

The mutual inductance M between stator and rotor windings can be expressed as

$$M = \frac{N_r \phi_{sr}}{I_s}$$

where N_r is the number of turns on the rotor winding, and ϕ_{sr} is the flux set up by the stator current which links the N_r turns of the rotor winding. Hence

$$T = I_s I_r \frac{dM}{d\theta} = I_s I_r \frac{N_r}{I_s} \frac{d\phi_{sr}}{d\theta}$$

$$= I_r N_r \frac{d\phi_{sr}}{d\theta}$$

and since $N_r = 1$ for the case considered

$$T = I_r \frac{d\phi_{sr}}{d\theta} \tag{2.10}$$

It remains to evaluate the quantity $d\phi_{sr}/d\theta$ from Fig. 2.10(b).

Let the rotor rotate by $d\theta$ in a clockwise direction. As the coil moves to the right in Fig. 2.10(b) it begins to link more downward pointing flux lines and fewer upward pointing flux lines. $d\phi_{sr}$ caused by the $d\theta$ rotation will therefore be negative.

If the coil side length is l then

$$d\phi_{sr} = -2\hat{B}l(Rd\theta)$$

where $lR\,d\theta$ is the area swept out by each coil side and $\hat{B}$ is the flux density over that area.

$$T = I_r \frac{d\phi_{sr}}{d\theta} = -I_r 2\hat{B}lR \tag{2.11}$$

and since $d\theta$ was taken as positive in a clockwise direction T is therefore directed anticlockwise.

The mutual inductance term in equation (2.8) is therefore equivalent to the *BlI* term used in the previous section. We should not find this surprising since both deal with the effects of the flux set up by one of the currents on the circuit of the other current.

The torque arising from the self-inductance terms in equation (2.8) is quite distinct from a *BlI* type torque and is worthy of further investigation.

2.7 RELUCTANCE TORQUE

In order to have non-zero self-inductance terms in equation (2.8) the self-inductance of one or both windings must vary with rotor position. A uniform air-gap machine does not give rise to such variation and we must consider a machine of the type shown in Fig. 2.11(a).

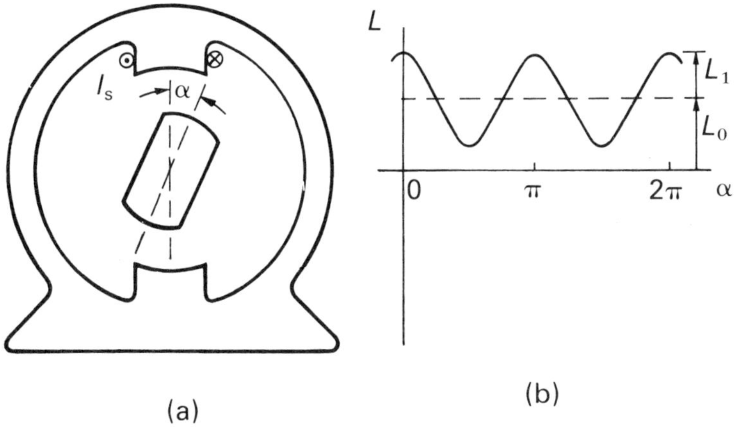

Fig. 2.11 (a) An elementary non-uniform air-gap machine and (b) the variation of self-inductance of the stator coil.

Fig. 2.11(b) shows how the self-inductance of the stator coil varies with the position of the rotor. L_s is a maximum when the reluctance of the flux path is a minimum, i.e. when $\alpha = 0$ and π. L_s is periodic in α and, neglecting higher harmonics, has the basic form

$$L_s = L_0 + L_1 \cos 2\alpha \tag{2.12}$$

If I_s is not a function of α then

$$\begin{aligned} T &= \tfrac{1}{2} I_s^2 \frac{dL_s}{d\alpha} \\ &= -\tfrac{1}{2} \cdot I_s^2 2L_1 \sin 2\alpha = -I_s^2 L_1 \sin 2\alpha \end{aligned} \tag{2.13}$$

Examination of equation (2.13) shows that for constant I_s the rotor will always attempt to get into the vertical position by the shortest possible route. This is the minimum reluctance position.

The torque produced by this mechanism is called a reluctance torque and reluctance motors will be considered in greater detail in Chapter 8. Reluctance torques are always present when the air-gap is non-uniform and there are windings on both rotor and stator.

2.8 THE ROTATING FIELD

When the rotor in Fig. 2.8(a) rotates at ω rad/s then the sinusoidal flux distribution produced by the rotor ampere-turn distribution will rotate with it. The flux distribution of Fig. 2.8(b) with one maximum and one minimum is referred to as a two-pole field.

In order that the rotor transfers power either to or from the stator it is necessary that the rotor experiences an average torque while rotating at ω. The stator winding currents must therefore produce a stator two-pole field distribution rotating at ω such that $T = K'\hat{B}_r\hat{B}_s \sin\beta$ has an average value. If the two fields are not rotating at the same speed then β will vary as one field sweeps past the other. Now the stator winding is stationary and a rotating field distribution can only be set up by a rotating current distribution. It follows that we must switch the stator currents from one part of the winding to another in order to produce the effect of a rotating current distribution. There are many ways of doing this but we will consider the most common situation, that of a three-phase winding carrying three-phase currents.

A three-phase stator winding has three separate coils each occupying one third of the stator surface next to the air-gap. Fig. 2.12 shows one such coil carrying current i_R. All the conductors carry the same current. The exact nature of the conductor interconnection will be considered later in Chapter 4.

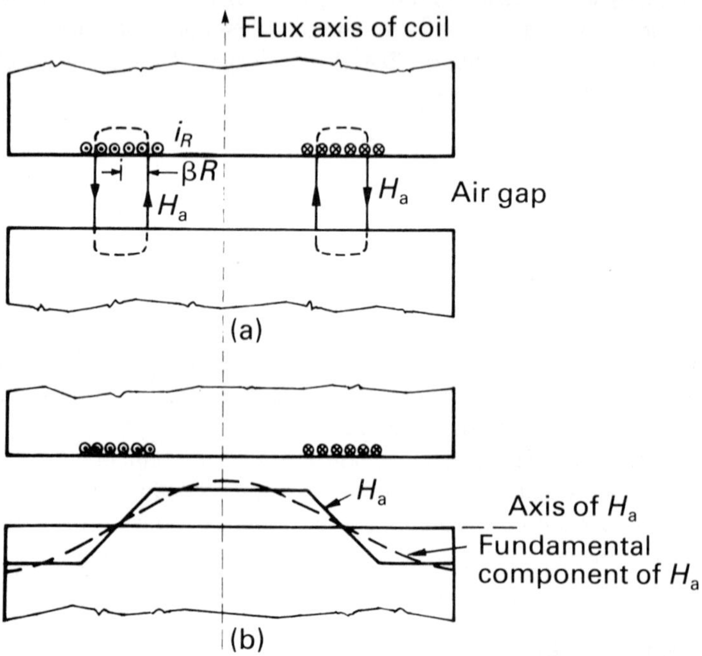

Fig. 2.12 (a) The stator current distribution for one-phase of a three-phase winding and (b) the air-gap flux density distribution set up by such currents.

If $i_R x$ is the ampere-turns per metre in the conductor band then for a flux loop which encloses part of the band, say βR as in Fig. 2.12(a), then

$H_a = i_R x \cdot \beta R/2g$. For a flux loop which encloses all of the conductor band

$$H_a = \frac{\pi R}{3} \cdot \frac{i_R x}{2g}$$

The corresponding values for B_a are given by $B_a = \mu_0 H_a$. Fig. 2.12(b) shows the flux density distribution in the air-gap. The fundamental component of the trapezoidal waveform is shown dotted and has amplitude $\hat{B}_r$ say. Let us concentrate on this fundamental component and neglect the other smaller harmonics for the moment. $\hat{B}_r$ is cosinusoidally distributed about the flux axis of coil R shown in Fig. 2.12(a), i.e. at some angle α from this axis the air-gap flux density is $\hat{B}_r \cos \alpha$.

Fig. 2.13 shows all three coils on the stator and their corresponding flux axes.

The bands of conductors are shown by a single conductor placed centrally in the band and the flux distribution corresponding to each coil is shown as a vector along the flux-axis of that coil. The projection of this vector on to a radial line gives the flux density in the air-gap at that line due to the coil in question.

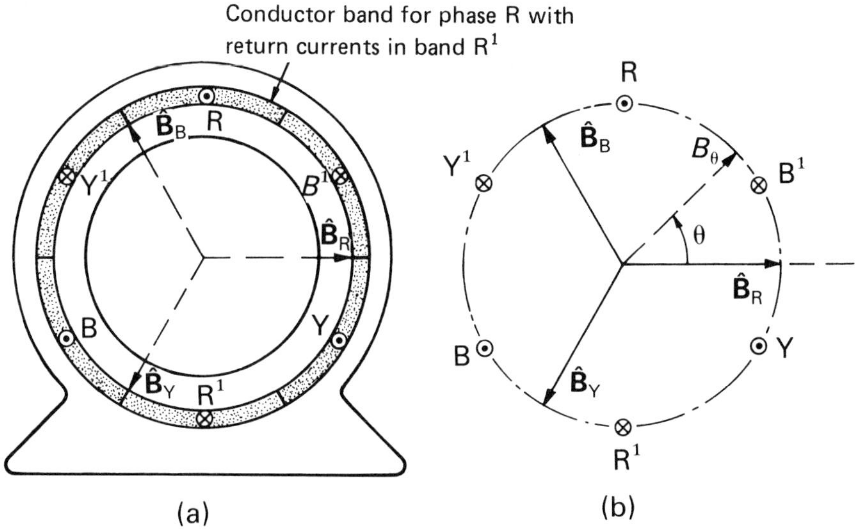

Fig. 2.13 (a) A three-phase stator current distribution and corresponding flux axes and (b) symbolic representation of such a winding.

Fig. 2.13(b) shows a symbolic representation of a three-phase stator winding and its corresponding air-gap flux density distribution which will be used extensively in later chapters. The conductor bands are shown by a single conductor at the centre of the band and the air-gap flux density distribution is shown as a vector along the flux axis.

The resultant air-gap flux density at a location θ away from the $\hat{B}_R$ axis is

$$B_\theta = \hat{B}_R \cos\theta + \hat{B}_Y \cos(120+\theta) + \hat{B}_B \cos(240+\theta) \tag{2.14}$$

Now if the currents i_R, i_Y, i_B form a three-phase set of phase sequence RBY then $i_R = \hat{I}\sin\omega t$, $i_B = \hat{I}\sin(\omega t - 120)$ and $i_Y = \hat{I}\sin(\omega t - 240)$. $\hat{B}_R$, $\hat{B}_Y$ and $\hat{B}_B$ will have amplitude variations corresponding to the amplitude variation in their coil currents such that

$$\hat{B}_R = \hat{B}\sin\omega t,\ \hat{B}_B = \hat{B}\sin(\omega t - 120) \quad \text{and} \quad \hat{B}_Y = \hat{B}\sin(\omega t - 240).$$

Substitution in equation (2.14) gives

$$B_\theta = \hat{B}\sin\omega t\cos\theta + \hat{B}\sin(\omega t - 240)\cos(240-\theta) + \hat{B}\sin(\omega t - 120)\cos(120-\theta) \tag{2.15}$$

By making use of the trigonometrical identity $\sin A\cos B = \frac{1}{2}\sin(A-B) + \frac{1}{2}\sin(A+B)$ and taking each term of equation (2.15) in turn

$$B_\theta = \tfrac{3}{2}\hat{B}\sin(\omega t - \theta) + \tfrac{1}{2}\hat{B}(\sin(\omega t + \theta) + \sin[(\omega t + \theta) - 120] + \sin[(\omega t + \theta) - 240]) \tag{2.16}$$

The second group of terms in equation (2.16) sums to zero since they form a symmetrical three-phase set hence

$$B_\theta = \tfrac{3}{2}\hat{B}\sin(\omega t - \theta) \tag{2.17}$$

At a fixed location in the air-gap (θ constant) the flux density has a sinusoidal time variation given by equation (2.17). A more important result, however, is observed by putting $\theta = \omega t - (\pi/2)$, i.e. the location or observer is going round in an anticlockwise direction of ω rad/s starting from a point $\pi/2$ radians behind the $\hat{B}_R$ axis at $t = 0$. Inserting the value of θ in equation (2.17) gives $B_{[\omega t - (\pi/2)]} = \frac{3}{2}\hat{B}$, i.e. the observer is going round on the crest of the resultant flux density wave. The observer sees a fixed value proving that the resultant wave is travelling at ω rad/s in the air-gap. The resultant flux density wave B_θ is cosinusoidally distributed about the observer position, e.g. put $\theta = \omega t - (\pi/2 \pm \alpha)$ in equation (2.17).

$$B_\theta = \tfrac{3}{2}\hat{B}\sin(\omega t - \omega t + \pi/2 \pm \alpha) = \tfrac{3}{2}\hat{B}\cos\alpha$$

When $i_R = \hat{I}$, $B_R = \hat{B}$, $\omega t = \pi/2$ then the crest of the resultant flux density wave given by $\theta = \omega t - (\pi/2)$ is in the location $\theta = 0$, i.e. aligned with the $\hat{B}_R$ axis. When $i_B = \hat{I}$, $B_B = \hat{B}$, $\omega t = (\pi/2) + 120$ then $\theta = \omega t - (\pi/2) = 120°$, i.e. aligned with $\hat{B}_B$. Similarly when $i_Y = \hat{I}$, θ can be shown to be aligned with $\hat{B}_Y$. Hence the flux axis of the resultant air-gap flux density wave is always aligned with the flux axis of a coil when that coil is carrying maximum current.

This analysis has shown how time varying currents applied to coils which are geometrically distributed on the stator can give rise to a rotating flux density distribution. This flux density distribution can now interact with the rotor derived flux distribution to produce an average torque when the rotor is going round at ω in an anticlockwise direction. The higher space harmonics which were neglected at the beginning of the analysis produce no average interaction with the rotor sinusoidal flux wave.

2.9 IRON LOSSES

The B/H curves of Fig. 2.5 do not give a completely accurate relationship for the behaviour of ferromagnetic material when subjected to an alternating m.m.f. They give an average relationship which serves as a design basis for magnetic circuits.

The iron toroid shown in Fig. 2.14(a) is subjected to an m.m.f. $\hat{I}N \sin \omega t$.

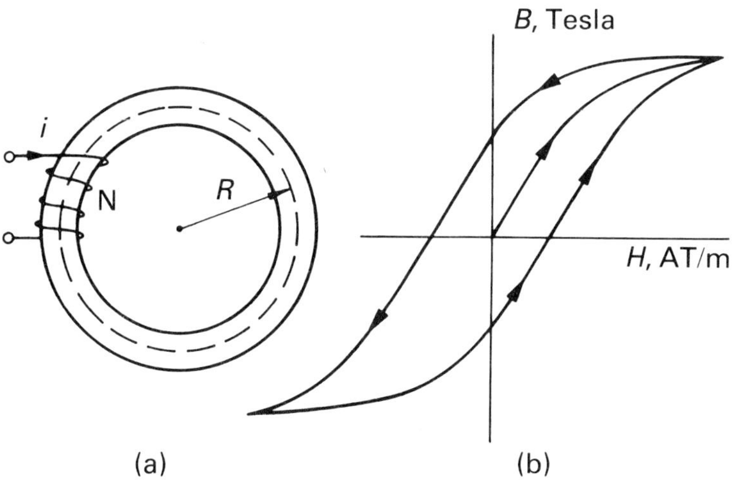

Fig. 2.14 (a) An iron toroid with a winding carrying a sinusoidal current and (b) the B/H loop in the iron.

H in the iron will be m.m.f./path length, hence $H = \hat{I}N \sin \omega t/2\pi R$. As i builds up from zero for the first time the path in the B/H plane of Fig. 2.13(b) starts from the origin and follows the average B/H curve until i reaches $\hat{I}$. As i decreases the B/H path departs from the average characteristic and when i reaches zero ($H = 0$) there is still some flux density in the iron. This is called the remanent flux or residual flux density. As i completes its sinusoidal cycle the B/H path forms a loop called a B/H or hysteresis loop. The extremities of this loop lie on the average characteristic but as with any mechanism exhibiting hysteresis, energy has been lost in the material in the process of generating the loop. A simple explanation of where the energy goes is in the alignment and realignment of the magnetic domains within the material. The alignment process encounters resistance due to inhomogeneities in the material.

Let the e.m.f. driving i through the coil in Fig. 2.14(a) be e and let the coil have resistance R and self-inductance L

$$\therefore \quad e = IR + L\frac{di}{dt} \tag{2.18}$$

but $L = N\phi/i$, where ϕ is the flux linking the coil.

$$\therefore \quad e = iR + N\frac{d\phi}{dt} \tag{2.19}$$

Multiply through equation (2.19) by i to get

$$ei = i^2R + iN\frac{d\phi}{dt} \tag{2.20}$$

ei is the instantaneous power input, i^2R is the power dissipated as heat in the resistor and $i \cdot N\,d\phi/dt$ is the rate of conversion of energy to stored magnetic energy.
But $\phi = B \times$ toroid cross-section area

and $$H = \frac{iN}{2\pi R},$$

hence

$$iN\frac{d\phi}{dt} = 2\pi R \times \text{area} \times H\frac{dB}{dt}$$

$$= (\text{toroid volume})\, H\frac{dB}{dt} \tag{2.21}$$

Hence the energy stored per unit volume in time dt is

$$H \cdot \frac{dB}{dt} \cdot dt = H \cdot dB$$

and the energy not recovered from the field in completing one B/H loop is

$$\oint H\,dB = \text{area of } B/H \text{ loop}$$

This loss is referred to as hysteresis loss.

Steinmetz found an empirical expression for the area of the B/H loop such that

$$\text{Area} = \eta B_m^x$$

where η is a coefficient depending on the material, B_m is the maximum flux density for the loop and x is an index again depending on the material. For many materials used in power devices $x \doteqdot 1.6$.

The power loss per unit volume is then

$$P_H = \eta B_m^{1.6} \cdot f \qquad (2.22)$$

where f is the frequency of the current i.

There is a second more familiar type of loss incurred when iron is subjected to an alternating m.m.f. Iron is itself a reasonable electrical conductor and when a block of iron is linked by a changing flux this gives rise to eddy currents in the iron. Fig. 2.15 shows a typical current path in the iron.

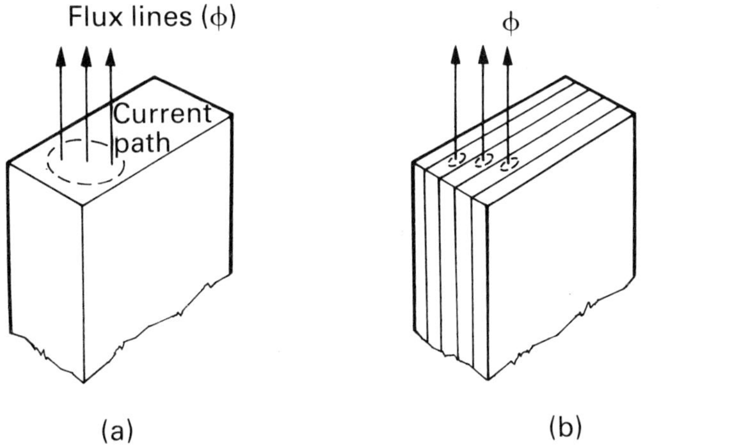

Fig. 2.15 Current path in (a) a solid iron core and (b) a laminated iron core.

The e.m.f. round the loop $e = d\phi/dt$, where $\phi = B \times$ loop area. The power loss due to this loop $= e^2/r$, where r is the loop resistance. When ϕ is an alternating flux of frequency f then the average power loss $= E_{rms}^2/r$. But

$$E_{rms} = \frac{\omega\hat{\phi}}{\sqrt{2}} = \sqrt{2}\pi f\hat{\phi}$$

An exact solution for the eddy current paths in an iron block is a complicated field problem. Suffice to say that eddy current loss for a given specimen is proportional to $f^2B_m^2$.

In order to minimise eddy current losses without increasing reluctance to the main flux, iron cores in power apparatus are laminated in thin strips. Fig. 2.15(b) shows the same flux ϕ as for Fig. 2.14(a) but current cannot flow across the interlamination insulation and the eddy currents are confined to much smaller loops inside a lamination. The loss is approximately proportional to t^2, where t is the lamination thickness.

The total iron loss for a given iron specimen can be expressed as

$$P_{\mathrm{I}} = P_{\mathrm{H}} + P_{\mathrm{E}} = \alpha f B_{\mathrm{m}}^{1.6} + \beta f^2 B_{\mathrm{m}}^2 \tag{2.23}$$

The coefficients α and β can be found from tests conducted at different frequencies and flux densities.

2.10 CONCLUSION

This chapter has dealt with the field theory needed to understand the steady state behaviour of electrical machines. Magnetic circuits and the relationship between current and field distributions in uniform air-gap machines have been covered in greater detail than can be found in most texts on e.m. fields. The relationship between the energy approach and the *Bll* approach to torque derivation has been explained and finally an expression to cover iron losses in magnetic circuits has been presented.

PROBLEMS

1 A 1000 turn coil is uniformly wound on an iron toroid of cross-sectional area $0.16 \times 10^{-4}\,\mathrm{m}^2$. The mean flux path length in the toroid is 0.5 m and it may be assumed that H and B are uniform across the toroid section. Neglecting leakage calculate the reluctance of the toroid flux path if μ_{r} for the iron is 800. Hence or otherwise calculate the self-inductance of the toroid winding.

Ans. 31×10^6 AT/Wb, 32.3 mH.

2 A magnetic circuit has a 4 mm air-gap cut in an iron toroid of cross-sectional area $36\,\mathrm{cm}^2$ and mean iron path length of 60 cm. A 200 turn coil wound on the toroid carries a steady current of 8 A. If the iron material is transformer sheet steel calculate the air-gap flux density neglecting leakage and fringing.

Ans. 0.47 T.

3 The elementary machine shown in Fig. 2.16 has a uniform air-gap. The mutual inductance between the rotor and stator coil is $M_0 \cos\theta$ where θ is the angle between the rotor and stator flux axes. The stator and rotor coil self-inductances are L_{s} and L_{r} respectively. The stator coil carries a direct current of I_{s} and the rotor coil carries a current $i_{\mathrm{r}} = \hat{I} \sin \omega t$.

(a) Derive an expression for the torque acting on the rotor when it is stationary at angle θ.

(b) Derive the torque expression when rotation takes place at Ω rad/s.

(c) At what speed if any is there an average torque?

Ans. (c) ω.

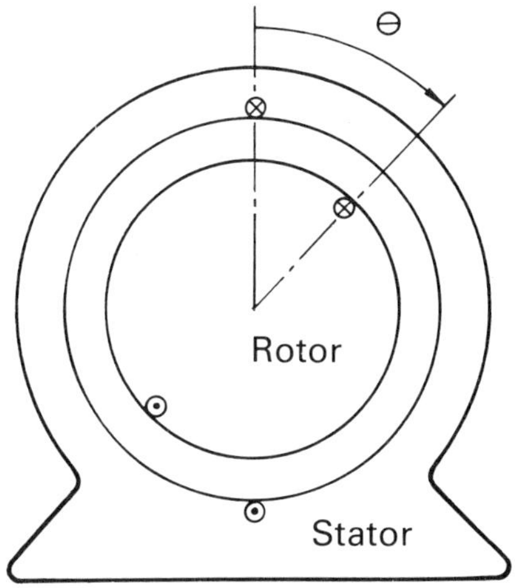

Fig. 2.16 An elementary uniform air-gap machine.

4 The area of the *BH* loop for a steel specimen taken to a maximum flux density of 1 T is 5.45 cm^2. The scale for *B* is 1 cm = 0.2 T and for *H* 1 cm = 160 AT/m. If the density for the specimen is 7.67 gm/cc calculate the hysteresis loss in watts/kg for the above conditions at 50 Hz.

Ans. 1.14.

5 A mild steel bar is bent into a loop so as to leave a uniform gap between its two plane end faces. Five centimetres of a straight rigid conductor lies in the air-gap parallel to the end faces. The conductor is connected in series with a 500 turn coil wound on the bar. The magnetic path has a mean length of 30 cm in the steel and $2\frac{1}{2}$ mm in the air-gap. Neglecting leakage and fringing estimate the coil current when the conductor experiences a force of 0.2 N.

6 A certain iron specimen in the shape of a toroid is tested for loss at various frequencies by applying a sinusoidal voltage to a coil wound over the toroid. The coil resistance is negligible. At 150 V, 30 Hz the total iron loss is 30 W and at 250 V, 50 Hz it is $63\frac{1}{3}$ W. Calculate the hysteresis and eddy current losses separately at 50 Hz.

Ans. 30 W, $33\frac{1}{3}$ W.

3

Transformers

3.1 INTRODUCTION

Alternating current power systems employ many different voltage levels. Large generators operate at approximately 20 kV, overhead transmission lines have operating levels from 3.3 kV up to 750 kV and domestic consumers draw power at approximately 200 V. The link between the various voltage levels is the power transformer.

This chapter will develop a simple equivalent circuit for the power transformer which is adequate to describe transformer steady state behaviour. Single and three-phase transformers will be covered and the analysis will draw heavily on the material of Chapter 1.

3.2 DEVELOPMENT OF EQUIVALENT CIRCUIT

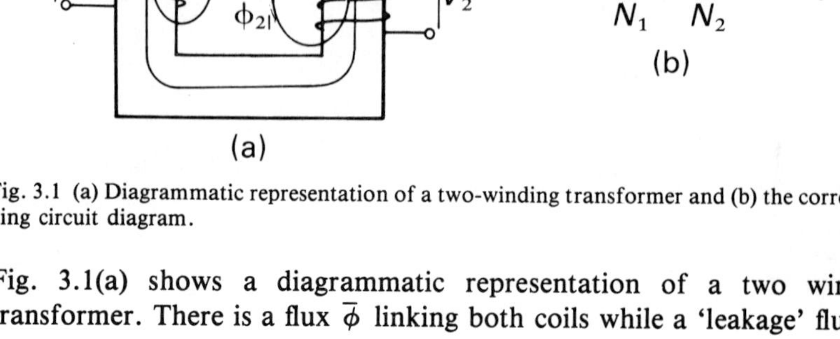

Fig. 3.1 (a) Diagrammatic representation of a two-winding transformer and (b) the corresponding circuit diagram.

Fig. 3.1(a) shows a diagrammatic representation of a two winding transformer. There is a flux $\bar{\phi}$ linking both coils while a 'leakage' flux $\bar{\phi}_{1l}$

or $\bar{\phi}_{2l}$ links only one coil. All variables are expressed as phasors since we are only concerned with steady state a.c. behaviour. This flux picture is not exact, e.g. not all of the leakage flux will link all the turns of a given coil, but it is a good approximation when the core material has a high μ_r and the coils are closely coupled. A practical transformer will tend to have one coil wound over the other on the same limb to ensure close coupling.

Fig. 3.1(b) shows the electrical circuit which corresponds to the system of Fig. 3.1(a). The circuit does not show any capacitive connections between the coils or between turns of a coil since such effects are negligible at power frequencies.

The circuit equations for the $\bar{V}_1$ and $\bar{V}_2$ loops are

$$\bar{V}_1 = L_1 \frac{d\bar{I}_1'}{dt} - M\frac{d\bar{I}_2}{dt} + \bar{I}_1' r_1 \tag{3.1}$$

$$-\bar{V}_2 = L_2 \frac{d\bar{I}_2}{dt} - M\frac{d\bar{I}_1'}{dt} + \bar{I}_2 r_2 \tag{3.2}$$

The relative sign of the self and mutual inductance voltage terms must be consistent in the two equations. The negative sign has been chosen to reinforce the fact that the fluxes set up by the primary (N_1) and the secondary (N_2) windings are in opposition when $\bar{V}_1$ is a power source and $\bar{V}_2$ is a power sink.

Equations (3.1) and (3.2) are not in a convenient form for making a quick assessment of the ratio V_1/V_2 at any current or of the voltage drop across the transformer when carrying current. A better form of relationships can be found if we make use of the fluxes indicated in Fig. 3.1(a).

Let $\bar{I}_1' N_1$ produce a flux $\bar{\phi}_1$ of which $\bar{\phi}_{12}$ links N_2 and $\bar{\phi}_{1l}$ does not link N_2: let $\bar{I}_2 N_2$ produce a flux $\bar{\phi}_2$ of which $\bar{\phi}_{21}$ links N_1 and $\bar{\phi}_{2l}$ does not link N_1: all of $\bar{\phi}_1$ links N_1 and all of $\bar{\phi}_2$ links N_2.

Hence

$$\bar{\phi}_1 = \bar{\phi}_{12} + \bar{\phi}_{1l} \quad \text{and} \quad \bar{\phi}_2 = \phi_{21} + \bar{\phi}_{2l} \tag{3.3}$$

In terms of inductances

$$\begin{aligned} L_1\bar{I}_1' &= N_1\bar{\phi}_1, & L_2\bar{I}_2 &= N_2\bar{\phi}_2 \\ M\bar{I}_1' &= N_2\bar{\phi}_{12}, & M\bar{I}_2 &= N_1\bar{\phi}_{21} \\ \bar{I}_1' l_1 &= N_1(\bar{\phi}_1 - \bar{\phi}_{12}), & \bar{I}_2 l_2 &= N_2(\bar{\phi}_2 - \bar{\phi}_{21}) \end{aligned} \tag{3.4}$$

where l_1 and l_2 are respectively the 'leakage' inductances of coils 1 and 2.

Equations (3.1) can now be rewritten making use of the relationships in equation (3.4) such that

$$\bar{V}_1 = N_1 \frac{d\bar{\phi}_1}{dt} - N_1\frac{d\bar{\phi}_{21}}{dt} + \bar{I}_1' r_1$$

$$= N_1 \frac{\mathrm{d}}{\mathrm{d}t}(\bar{\phi}_1 - \bar{\phi}_{21}) + \bar{I}_1' r_1$$

$$= N_1 \frac{\mathrm{d}}{\mathrm{d}t}(\bar{\phi}_1 - \bar{\phi}_{12} + \bar{\phi}_{12} - \bar{\phi}_{21}) + \bar{I}_1' r_1$$

$$= N_1 \frac{\mathrm{d}}{\mathrm{d}t}(\bar{\phi}_1 - \bar{\phi}_{12}) + N_1 \frac{\mathrm{d}}{\mathrm{d}t}(\bar{\phi}_{12} - \bar{\phi}_{21}) + \bar{I}_1' r_1$$

$$= l_1 \frac{\mathrm{d}\bar{I}_1'}{\mathrm{d}t} + N_1 \frac{\mathrm{d}\bar{\phi}}{\mathrm{d}t} + \bar{I}_1' r_1 \qquad (3.5)$$

where $\bar{\phi}$ is the net linking flux in the core given by

$$\bar{\phi} = \bar{\phi}_{12} - \bar{\phi}_{21}$$

By a similar process it can be shown that equation (3.2) becomes

$$-\bar{V}_2 = l_2 \frac{\mathrm{d}\bar{I}_2}{\mathrm{d}t} - N_2 \frac{\mathrm{d}\bar{\phi}}{\mathrm{d}t} + \bar{I}_2 r_2 \qquad (3.6)$$

If we further define

$$\bar{E}_1 = N_1 \frac{\mathrm{d}\bar{\phi}}{\mathrm{d}t}, \; \bar{E}_2 = N_2 \frac{\mathrm{d}\bar{\phi}}{\mathrm{d}t}$$

then the circuit diagram of Fig. 3.1(b) can be redrawn as in Fig. 3.2.

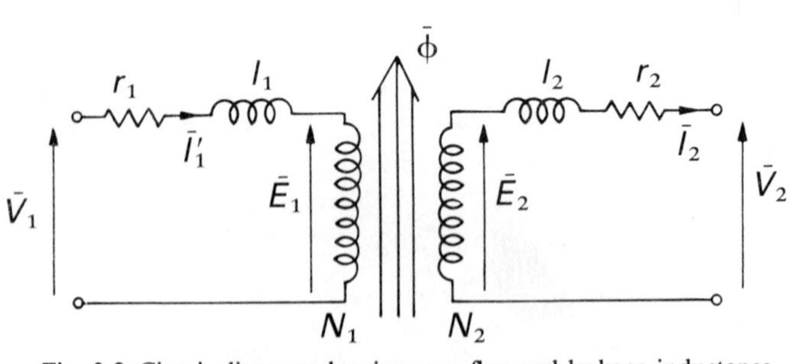

Fig. 3.2 Circuit diagram showing core flux and leakage inductance.

The core flux path will have a small reluctance and to produce $\bar{\phi}$ will require an m.m.f. The m.m.f. comes from the difference in ampere turns of the two windings with the primary (V_1) supply winding producing the larger m.m.f. If the core reluctance is $\mathcal{R}_c$ then

$$\bar{I}_1' N_1 - \bar{I}_2 N_2 = \bar{\phi} \times \mathcal{R}_c$$

Let us define a component of the primary current $\bar{I}_1'$ as $\bar{I}_m$, the magnetising current, such that

$$\bar{I}_m N_1 = \bar{I}_1' N_1 - \bar{I}_2 N_2 = \bar{\phi} \times \mathcal{R}_c \tag{3.7}$$

$$\therefore \quad (\bar{I}_1' - \bar{I}_m) N_1 = \bar{I}_2 N_2, \text{ and if } \bar{I}_1 = \bar{I}_1' - \bar{I}_m$$

then

$$\bar{I}_1 N_1 = \bar{I}_2 N_2 \tag{3.8}$$

Making use of equations (3.7) and (3.8) we can redraw the circuit of Fig. 3.2 to give Fig. 3.3.

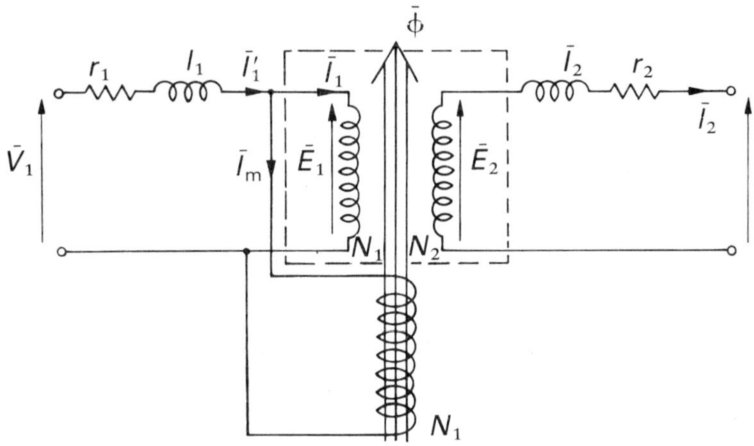

Fig. 3.3 Circuit diagram showing magnetising ampere-turns separate from 'perfect' transformer.

The magnetising ampere-turns have been shown separately leaving us with two coils of N_1 and N_2 turns which link the flux $\bar{\phi}$ but do not contribute to its production since $\bar{I}_1 N_1 - \bar{I}_2 N_2 = 0$. The part of the circuit inside the dotted contour on Fig. 3.3 is often referred to as an 'ideal' transformer. All of the deficiencies of the real transformer have been accounted for by elements placed outside the box leaving us with two perfectly coupled coils for which

$$\frac{\bar{E}_1}{\bar{E}_2} = \frac{N_1}{N_2} \quad \text{and} \quad \frac{\bar{I}_1}{\bar{I}_2} = \frac{N_2}{N_1}$$

When operating at power frequency the ratio of $\bar{E}_1/\bar{I}_m = jX_0$, the magnetising reactance. X_0 is placed in parallel with $\bar{E}_1$ as will be seen in Fig. 3.4.

There is one further deficiency of the real transformer which is not yet accounted for in Fig. 3.3. $\bar{\phi}$ is an a.c. flux and we will therefore generate hysteresis and eddy current losses in the core material. The core will of course be laminated in order to minimise the latter. In an equivalent circuit losses must entail a resistor, say R_0. The question is where to put it. In Section 2.9 we saw that iron loss for a given specimen was given by

$$P_I = \alpha f B_m^{1.6} + \beta f^2 B_m^2$$

Power transformers are operated at fixed frequency and this expression can be modified to

$$P_I = \alpha' \phi_m^{1.6} + \beta' \phi_m^2$$

$$\simeq \gamma \phi_m^2 \tag{3.9}$$

But $E_1 = N_1 \, d\phi/dt = N_1 \omega \phi_m \cos \omega t$ for sinusoidal ϕ, and hence for fixed frequency E_1 (r.m.s.) $= K\phi_m$.

Therefore, from equation (3.9) we can say

$$P_I \simeq \gamma' E_1^2 \tag{3.10}$$

To get a loss proportional to E_1^2 we must put R_0 in parallel with $\bar{E}_1$. The equivalent circuit now becomes as in Fig. 3.4(a) with R_0, X_0 representing the core losses and magnetising reactance. $|\bar{I}_0| = |\bar{I}_m + \bar{I}_w|$ is typically $I_1/20$ when the transformer is running at rated current. The voltage drop across r_1 and $x_1(\omega l_1)$ is typically 2 or 3 per cent of V_1 under these conditions. One final approximation therefore shifts the core components R_0, X_0 to the supply terminals as in Fig. 3.4(b) without incurring any significant loss of accuracy over and above that which has already been lost due to our other approximations.

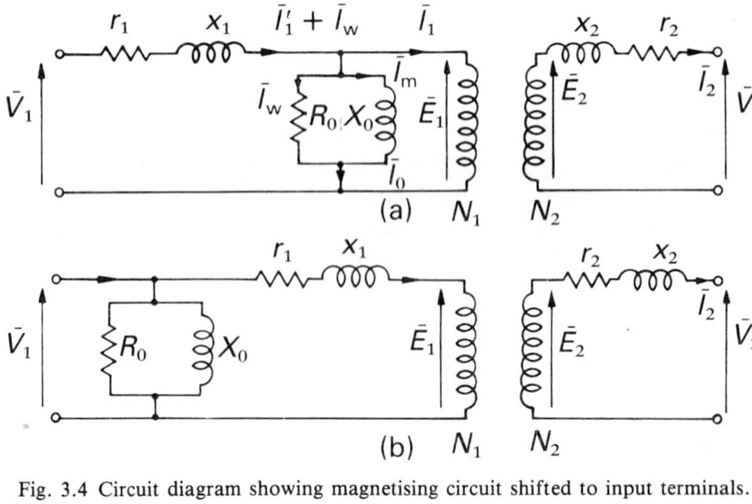

Fig. 3.4 Circuit diagram showing magnetising circuit shifted to input terminals.

The position of the ideal transformer in Fig. 3.4(b) may be altered provided the impedance values are altered in accordance with this change in position. The rules governing the impedance modification will now be derived.

From Fig. 3.4(b) the circuit equations may be written

$$\bar{V}_1 = \bar{I}_1(r_1 + jx_1) + \bar{E}_1 \tag{3.11}$$

$$\bar{E}_2 = \bar{I}_1(r_2 + jx_2) + \bar{V}_2 \tag{3.12}$$

Substitute for $\bar{E}_2$ and $\bar{I}_2$ in equation (3.12) in terms of $\bar{E}_1$, $\bar{I}_1$ and $n = N_2/N_1$ the turns ratio. Equation (3.12) becomes

$$n\bar{E}_1 = \frac{\bar{I}_1}{n}(r_2 + jx_2) + \bar{V}_2$$

$$\bar{E}_1 = \bar{I}_1\left(\frac{r_2 + jx_2}{n^2}\right) + \frac{\bar{V}_2}{n} \tag{3.13}$$

Substituting (3.13) into (3.11) gives

$$\bar{V}_1 = \bar{I}_1\left[(r_1 + jx_1) + \frac{1}{n^2}(r_2 + jx_2)\right] + \frac{\bar{V}_2}{n} \tag{3.14}$$

The circuit has been redrawn in Fig. 3.5(a) where it can be seen that the secondary impedances have been shifted to the primary side and modified by the factor $1/n^2$. The ideal transformer has been moved to the load terminals (V_2). If we had chosen to eliminate $\bar{E}_1$ and $\bar{I}_1$ then we would have arrived at the circuit shown in Fig. 3.5(b) with the ideal transformer moved to the supply terminals. In this case transferred impedances are multiplied by n^2. Note that watts and vars, I^2r and I^2x, are conserved in this transfer process. This latter point helps to eliminate confusion about whether an

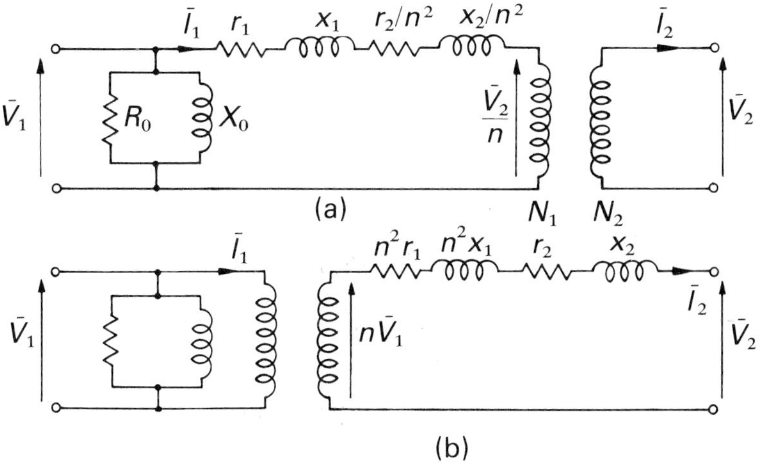

Fig. 3.5 Transfer of impedances into (a) the primary circuit and (b) the secondary circuit.

impedance goes up or down by n^2 when being transferred. If the current is less on the side to which the transfer is taking place then the impedance must increase.

The leakage reactances and winding resistances are lumped together into one combined impedance to give the final form of equivalent circuit shown in Fig. 3.6(a) and (b). Which of the two circuits, (a) or (b), is chosen to represent the transformer will depend on the way in which the information is presented.

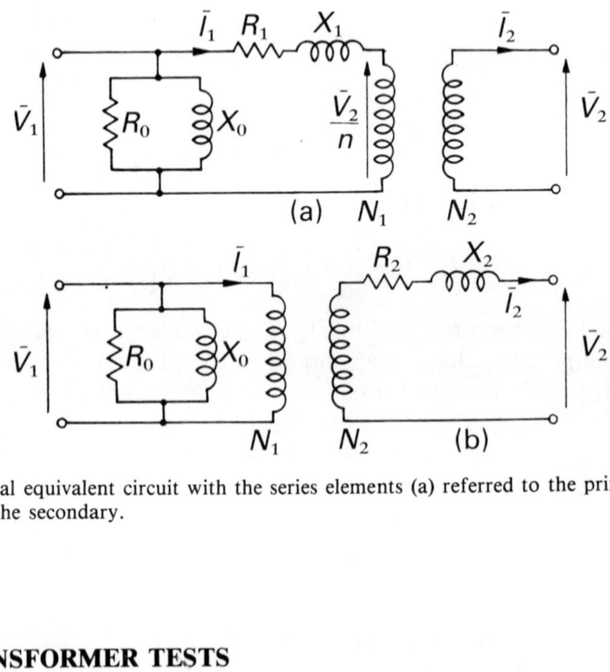

Fig. 3.6 Final equivalent circuit with the series elements (a) referred to the primary and (b) referred to the secondary.

3.3 TRANSFORMER TESTS

There are two standard tests performed on a transformer to determine the components of the equivalent circuit.

3.3.1 The open circuit test

One winding of the transformer is supplied at its rated voltage, V_1 say, and the other winding is on open circuit. The current in this latter winding, I_2 say, will be zero. Instruments are connected to read the supply voltage (V_1), current (I) and power (P_0) and the voltage across the open circuited winding (V_2). It is normal to refer to the winding connected to the supply as the primary. With reference to Fig. 3.6(a) with I_2 and hence I_1 set to zero,

$$n = \frac{V_2}{V_1}, \qquad Z_0 = \frac{V_1}{I}, \qquad R_0 = \frac{V_1^2}{P_0}, \qquad X_0 = \frac{V_1}{I \sin \theta}$$

where $$\theta = \cos^{-1}\left(\frac{P_0}{V_1 I}\right).$$

The turns ratio and the elements of the magnetising circuit referred to the primary are determined from this test. The approximation involved in shifting the magnetising circuit to the supply terminals ignores the copper loss ($I^2 r_1$) in the primary winding resistance. In a well designed power transformer (V_1^2/R_0), the iron loss at rated voltage, will be approximately equal to the copper loss at rated current ($I_1^2 R_1$). As noted earlier I(I_0 in this test) is about $\frac{1}{20}$th of the rated I_1 value and hence $I^2 R_1$ would be $\frac{1}{400}$ of the rated copper loss. $I^2 r_1$ is roughly half of $I^2 R_1$ and hence the copper loss in this test is extremely small in comparison to the iron loss.

3.3.2 The short circuit test

The secondary winding is short-circuited usually by heavy gauge conductor or copper bar in order to minimise the resistance of the short circuit. The primary winding is supplied at a much reduced voltage sufficient only to cause I to reach its rated value. A typical value for V_1 would again be of the order V_1 (rated)/20. Instruments are connected to read the supply voltage, current and power (P_{sc}). From these readings we obtain

$$R_1 = \frac{P_{sc}}{I^2}, \qquad Z_1 = \frac{V_1}{I_1}, \qquad X_1 = \sqrt{Z_1^2 - R_1^2}$$

I_0 is negligible in this test due to the reduced voltage and likewise iron loss is negligible in comparison to copper loss.

The short circuit test need not be carried out supplying the same winding as for the open circuit test. The rating of the test supplies available and the transformer rating and turns ratio will decide which winding to supply in a given test. The impedances calculated from test results are the impedances referred to the winding supplied during that test.

Example 3.1

A 2 kV/440 V, 200 kVA, 50 Hz transformer gave the following test results.

Open circuit test supplying the HV side:

$$V = 2\,\text{kV}, \qquad P_0 = 1.75\,\text{kW}, \qquad I = 1.8\,\text{A}$$

Short circuit test supplying the LV side:

$$V = 13\,\text{V}, \qquad P_{sc} = 1\,\text{kW}, \qquad I = 300\,\text{A}$$

Calculate the elements of the transformer equivalent circuit.

From the open circuit test:

$$\cos\theta = \frac{1.75\times10^3}{2\times10^3\times1.8} = 0.486,\ \sin\theta = 0.874$$

$$X_0 = \frac{V}{I\sin\theta} = \frac{2\times10^3}{1.8\times0.874} = \underline{1.27\,\text{k}\Omega}$$

$$R_0 = \frac{V}{I\cos\theta} = \frac{2\times10^3}{1.8\times0.486} = \underline{2.29\,\text{k}\Omega}$$

(Check this value against $R_0 = V^2/P_0$)

From the short circuit test:

$$Z = \frac{V}{I} = \frac{13}{300} = 0.043\ \Omega$$

$$R = \frac{P_{sc}}{I^2} = \frac{10^3}{300^2} = \underline{0.011\ \Omega}$$

$$X = \surd(0.043^2 - 0.011^2) = \underline{0.042\ \Omega}$$

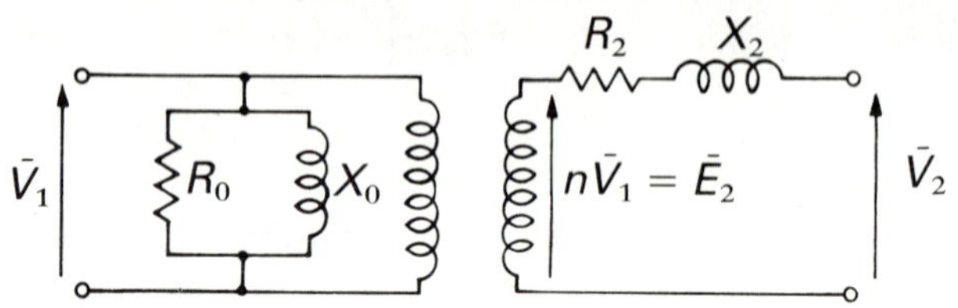

Fig. 3.7 Equivalent circuit for the transformer of Example 3.1.

Fig. 3.7 shows the equivalent circuit with the series elements on the LV side and the magnetising impedance on the HV side. If the transformer was to be supplied from the LV side then Z_0 would be divided by $(2\times10^3/440)^2$ and shifted to the LV terminals.

3.4 REGULATION AND EFFICIENCY

It is important that the output voltage of a power transformer remain sensibly constant over the range of currents supplied. Furthermore, the power loss in the device should be as small as possible. Both of these considerations require that the series elements of the equivalent circuit, R and X, should be small. Power loss is also affected by R_0 which should therefore be as large as possible.

The volt drop from no load to full load current is conveniently expressed as the percentage 'regulation' where

$$\text{regulation (\%)} = \frac{V_{\text{NL}} - V_{\text{FL}}}{V_{\text{NL}}} \times 100 \tag{3.15}$$

V_{NL} is the no load (current) output voltage and V_{FL} is the output voltage when the transformer is delivering full load current.

Example 3.2

For the transformer of Example 3.1 calculate the regulation and efficiency when the transformer supplies a 440 V, 200 kVA load at 0.8 p.f. lagging.

The load current

$$I_2 = \frac{200 \times 10^3}{440} = \underline{454.5\ \text{A}}$$

With reference to Fig. 3.7, and taking $\overline{V}_2$ as reference phasor,

$$\overline{I}_2 = 454.5 \angle -36.87^\circ, \ (\cos^{-1}(0.8) = 36.87)$$

$$\overline{E}_2 = \overline{V}_2 + \overline{I}_2\overline{Z}_2$$

$$= 440 \angle 0 + 454.5 \angle -36.87^\circ.\ (0.011 + j0.042)$$

$$= 440 \angle 0 + 19.73 \angle 38.45^\circ$$

$$= 455.45 + j12.27 = \underline{455.6 \angle 1.54^\circ}$$

$$\therefore \quad \text{regulation} = \left(\frac{455.6 - 440}{455.6}\right) \times 100 = \underline{3.4\%}$$

$$\left(\text{Note that the HV supply remains constant at } 455.6 \times \frac{2 \times 10^3}{440} = \underline{2.07\ \text{kV}}\right)$$

The iron loss is proportional to V_1^2 hence,

$$P_0 \text{ at } 2.07\ \text{kV} = \frac{1.75 \times 10^3 \times 2.07^2}{2^2} = \underline{1.87\ \text{kW}}$$

The 'copper' loss $= I_2^2R_2 = 454.5^2 \times 0.011 = \underline{2.27\ \text{kW}}$

$$\therefore \quad \text{Efficiency} = \frac{\text{Output}}{\text{Output} + \text{losses}} = \frac{200 \times 0.8}{200 \times 0.8 + 1.87 + 2.27}$$

$$= \underline{97.5\%}$$

Appendix B gives an alternative approximate way to calculate regulation when $V_2 \doteqdot E_2$ and the angle between $\overline{V}_2$ and $\overline{E}_2$ is very small.

$$\text{Regulation} = \frac{I_2R_2 \cos\theta - I_2X_2 \sin\theta}{V_2}$$

For a lagging power factor θ is negative, hence in this case

$$\text{Regulation} = \frac{454.5(0.011 \times 0.8 + 0.042 \times 0.6)}{440} \times 100$$

$$= \frac{454.5 \times 0.34}{440} \times 100$$

$$= \underline{3.5\%}$$

3.5 MAGNETISING CURRENT WAVESHAPE

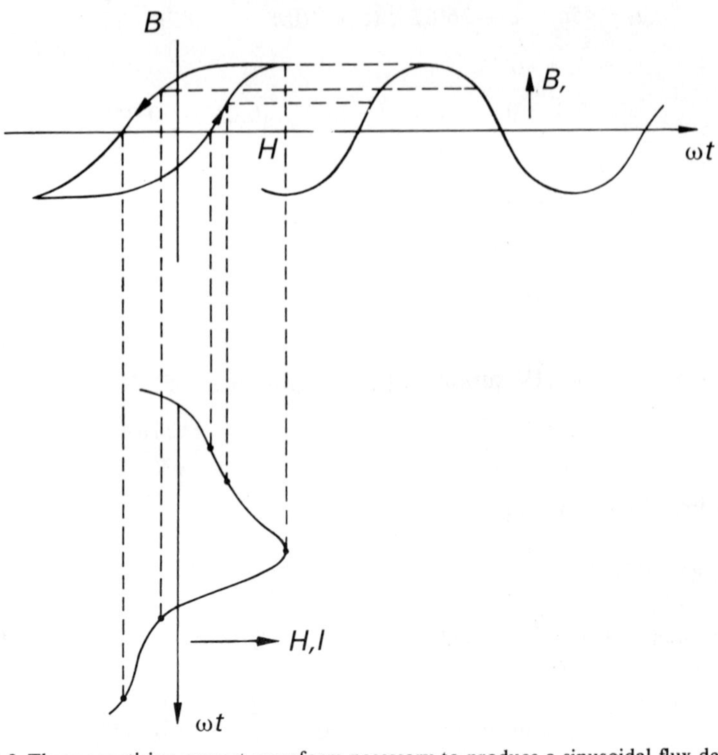

Fig. 3.8 The magnetising current waveform necessary to produce a sinusoidal flux density in the core.

When the supply voltage is sinusoidal then $\bar{\phi}$ must be cosinusoidal since $\bar{E}_1 = N_1 \mathrm{d}\bar{\phi}/\mathrm{d}t$ and $\bar{E}_1$ is very nearly equal to $\bar{V}_1$. The core flux density $\bar{B}$ must likewise be cosinusoidal. Fig. 3.8 shows how a flux density $\bar{B}$ of cosinusoidal waveshape produces, via the B/H loop for the core material, a non-sinusoidal H wave. Now $H = I_h N_1/l$, where I_h is the component of the magnetising current which accounts for the setting up of the hysteresis loop. It will not contain the eddy current component of the magnetising current. l is the mean length of the core magnetic path. I_h will therefore have the same waveshape as H in Fig. 3.8. I_h is composed of a fundamental component plus a number of odd harmonics. Fourier analysis indicates that the predominant harmonic is the third with lesser contributions from the fifth, seventh and higher odd harmonics.

The total magnetising current I_0 consists of $(I_h + I_e)$ where I_e is a sinusoidal component to account for the eddy current losses.

A power transformer on open circuit or light load will draw a non-sinusoidal current from the supply. Most power transformers, however, tend to operate close to full load (rated current) and since I_0 is typically 1/20th of rated current the harmonic content of the total current drawn from the supply is very small.

Fig. 3.8 indicates that $\hat{B}$ should not go much beyond the linear portion of the average B/H curve if sharp peaks are to be avoided in H and hence I_0. These peaks will increase the harmonic content of I_0. $\hat{B}$ in the region 1 to 1.2 T would be typical for transformer sheet steel. (see Fig. 2.5)

3.6 THE AUTOTRANSFORMER

In the two winding transformer $\bar{E}_1$ and $\bar{E}_2$ have been taken from separate windings on the same core.

The ratio $E_1/N_1 = E_2/N_2$, the volts/turn, determines the core cross-section, A_c. This is because $e_1/N_1 = \mathrm{d}\phi/\mathrm{d}t = A_c \mathrm{d}B/\mathrm{d}t$ and for sinusoidal flux $e_1/N_1 = A_c \omega \hat{B} \cos \omega t$. $\hat{B}$ for reasons explained in Section 3.5 must be limited to say 1.2 Tesla. Hence, for a 50 Hz system, $E_1/N_1 = A_c \omega B = A_c\ 1.2/\sqrt{2} \times 2\pi 50 = 266.6 A_c$ volts/turn. If we now consider the case where $N_2 < N_1$ then Fig. 3.9 shows an alternative way of deriving E_2 completely dispensing with the secondary winding N_2. The core cross-section is as before but the winding space taken up by the secondary winding is now available to allow an increase in the primary winding cross-sectional area thereby allowing it to carry more current. This will result in an increase in rating (VI) for the same transformer overall dimensions. It is evident that the greatest advantage is obtained when $N_2 \doteqdot N_1$. The disadvantage of this connection is the loss of isolation between the input and output circuits. This is particularly important in three-phase transformers where one may wish to use a different win-

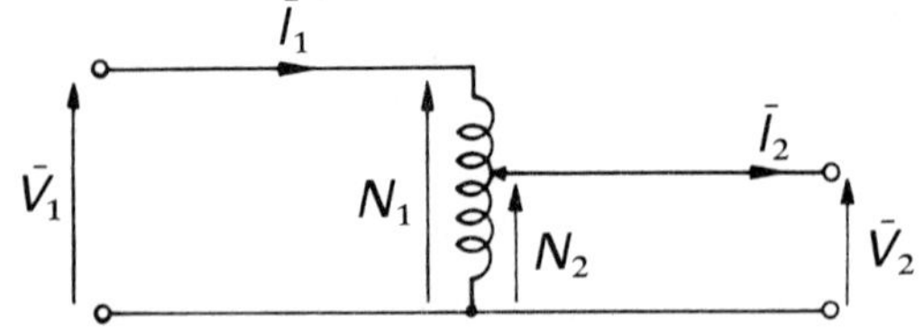

Fig. 3.9 The 'ideal' autotransformer circuit.

ding and earthing arrangement on each side of the transformer. The 'ideal' autotransformer equations are

$$\frac{\bar{V}_1}{\bar{V}_2}=\frac{N_1}{N_2},\quad \bar{V}_1\bar{I}_1^*=\bar{V}_2\bar{I}_2^*$$

hence

$$\frac{\bar{I}_2}{\bar{I}_1}=\frac{N_1}{N_2}$$

The current flowing in the N_2 turns of Fig. 3.9 is $(\bar{I}_1-\bar{I}_2)$. When N_2 and N_1 are equal then the transformer winding carries no current and when $N_2=N_1/2$ both the upper and lower half of the winding carry a current $\bar{I}_1$, but in opposite directions. If $N_2<(N_1/2)$ then the N_2 lower turns will carry a current in excess of I_1. Since the whole winding will have the same cross-section of conductor it is desirable to have no part of the winding carrying a current in excess of I_1.

Autotransformers are most often used when the voltage ratio is between 1 and 2 and when isolation between the circuits on opposite sides of the transformer is not required.

3.7 THREE-PHASE TRANSFORMERS

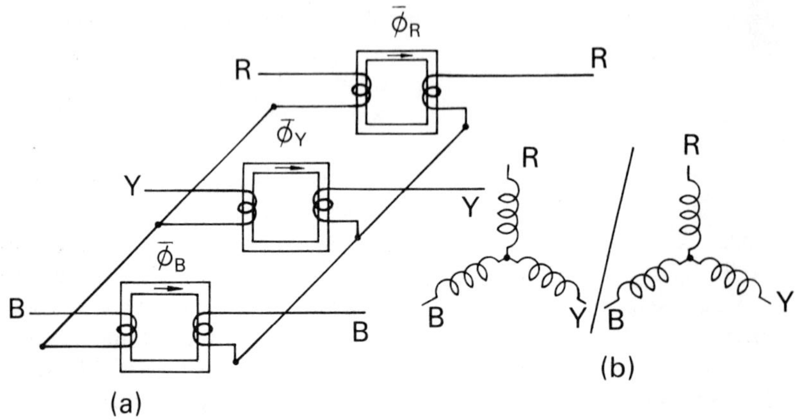

Fig. 3.10 (a) Three single-phase transformers connected to form a three-phase 'bank' and (b) corresponding circuit diagram.

Three single phase transformers can be connected together as in Fig. 3.10(a) to form a three-phase bank. Both windings are connnected in star and Fig. 3.10(b) shows how the connection of Fig. 3.10(a) is portrayed in circuit diagrams.

Use can be made of the fact that the core fluxes $\bar{\phi}_R$, $\bar{\phi}_Y$ and $\bar{\phi}_B$ form a three-phase symmetrical set which therefore sum to zero. This allows a core arrangement of the type shown in Fig. 3.11 to be used which effects considerable saving in core material compared to Fig. 3.10(a).

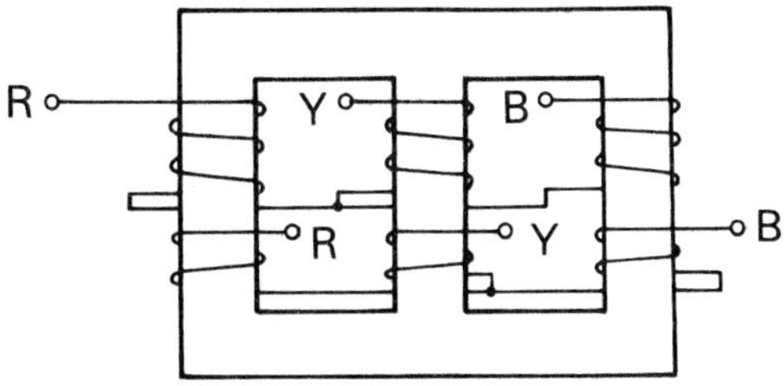

Fig. 3.11 A 'core' type three-phase transformer.

3.7.1 Three-phase transformer connections

There are three possible configurations for a three-phase transformer. Both windings may be connected in star as in Fig. 3.10(b), both windings may be connected in delta as in Fig. 3.12(a) or one winding can be connected in delta and the other in star as in Fig. 3.12(b).

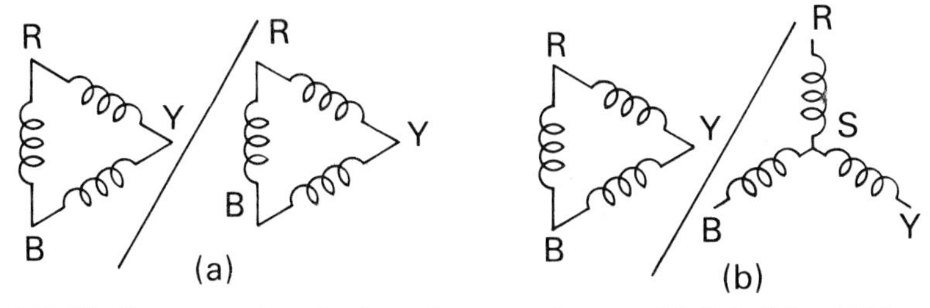

Fig. 3.12 Winding connections in three-phase transformers. (a) Delta/delta (Δ/Δ) or (b) delta/star (Δ/λ).

In Section 3.5 it was shown that the magnetising current contained a component at fundamental frequency with a significant component at three times fundamental frequency. If the voltage across each phase is to remain sinusoidal then the phase magnetising currents must be of the following form

$$I_{R0} = I_1 \sin \omega t + I_3 \sin 3\omega t$$

$$I_{Y0} = I_1 \sin (\omega t - 120) + I_3 \sin 3(\omega t - 120)$$

$$= I_1 \sin (\omega t - 120) + I_3 \sin 3\,\omega t$$

$$I_{B0} = I_1 \sin(\omega t - 240) + I_3 \sin 3(\omega t - 240)$$
$$= I_1 \sin(\omega t - 240) + I_3 \sin 3\,\omega t$$

For the star connection the three currents sum at the star point to give $3I_3 \sin 3\,\omega t$. A neutral connection must be provided if the third harmonic is not to be forced to zero thereby distorting the flux and voltage waveshape. On the other hand the line current into either of the delta connections of Fig. 3.12(a) or (b) will consist of the difference of two phase currents so that the third harmonic components of the phase currents will cancel out in the line. The third harmonic component circulates in the delta and does not appear in the line currents. A delta connection therefore allows a sinusoidal flux and voltage with no third harmonic currents in the supply line. For this reason the vast majority of three-phase transformers have a delta connected winding and in cases where it is not convenient to have either the primary or secondary connected in delta a tertiary winding is added and connected in delta. The tertiary winding may or may not be used to supply an additional load but it serves to carry the circulating third harmonic current required by the sinusoidal flux in each limb of the core.

When the primary and secondary windings have the same configuration then any balanced three-phase problem can be dealt with by considering one phase only. The currents and voltages in the other phases can then be derived from the three-phase symmetry. Care must be exercised in breaking down test and load information into the relevant quantities for dealing with one phase. For instance, if three of the transformers used in Example 3.1 were connected in a delta/delta configuration the test information would be given as follows:

Open circuit test supplying the HV side

$$V_L = 2\,\text{kV}, \qquad P_0 = 5.25\,\text{kW}, \qquad I_L = 3.12\,\text{A}$$

Short circuit test supplying the LV side

$$V_L = 13\,\text{V}, \qquad P_{sc} = 3\,\text{kW}, \qquad I_L = 519.6\,\text{A}$$

Since it is a delta connection on both windings the phase voltage equals the line voltage, the phase current is $1/\sqrt{3}$ times the line current and the phase power is 1/3 times the total power. The equivalent circuit per phase is then as derived in Example 3.1.

The delta/star connection is much more difficult to deal with and is best considered in the context of an example

Example 3.4
A 12 kVA, 415 V/5.72 kV, delta/star three-phase transformer gave the following test results.

Open circuit test on LV winding;

$$V_L = 415\text{ V}, \qquad P_0 = 120\text{ W}, \qquad I_L = 0.8\text{ A}$$

Short circuit test on HV winding;

$$V_L = 210.1\text{ V}, \qquad P_{sc} = 150\text{ W}, \qquad I_L = 1.16\text{ A}$$

The transformer supplies a three-phase load of 9.6 kW at 0.8 p.f. lagging at a line voltage of 5.72 kV.

Calculate the line voltage and line current on the LV side and the reading on two wattmeters connected to read input power. Calculate also the efficiency and regulation of the transformer.

Since we are not asked for the elements of the equivalent circuit there is no need to calculate R_0 and X_0. We will need P_0 for the efficiency calculation and we need to know the phase angle of I_L in the open circuit test in order to add it to the load current to achieve the total input current.

Open circuit test
Consider one phase of delta winding

$$V_L = V_p = 415, \qquad I_p = \frac{I_L}{\sqrt{3}} = 0.462\text{ A}, \qquad P_p = 40\text{ W}, \qquad I_p = I_0$$

$$\therefore \quad \cos\theta_0 = \frac{40}{415 \times 0.462} = 0.209, \qquad \theta_0 = 77.96^\circ$$

Because the magnetising circuit is inductive I_0 will lag V_L by θ_0.

Short circuit test
Consider one phase of the star winding

$$V_p = \frac{V_L}{\sqrt{3}} = 121.3\text{ V}, \qquad I_p = I_L = 1.16\text{ A}, \qquad P_p = 50\text{ W}.$$

$$\therefore \quad R = \frac{50}{1.16^2} = \underline{37.16\ \Omega}$$

$$Z = 121.3/1.16 = 104.57\ \Omega$$

$$\therefore \quad X = \sqrt{(Z^2 - R^2)} = 97.74\ \Omega$$

Load condition

The load is on the star side so consider one phase of the secondary winding plus the load in that phase.

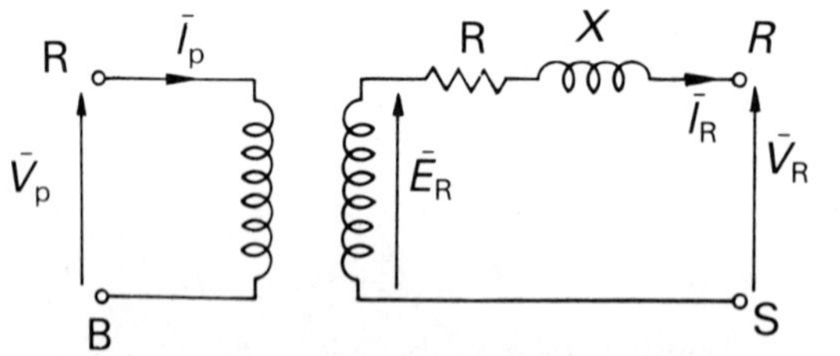

Fig. 3.13 Equivalent circuit for one-phase referred to the secondary winding.

Take $\bar{V}_R = 5.72 \times 10^3/\sqrt{3} \angle 0°$ as reference phasor, that is $\bar{V}_R = 3.3 \times 10^3 \angle 0°$.

$\bar{I}_R$ is derived from $P = \sqrt{3} V_L I_L \cos\theta_L$ (Note: $\cos\theta_L$ is load power factor.)

$$\therefore \quad I_R = \frac{9.6 \times 10^3}{5.72 \times 10^3 \times \sqrt{3} \times 0.8} = 1.21 \text{ A}$$

$$\therefore \quad \bar{I}_R = 1.21 \angle -36.87°$$

From Fig. 3.13

$$\bar{E}_R = \bar{V}_R + \bar{I}_R \cdot (R + jX)$$

$$= 3.3 \times 10^3 \angle 0° + 1.21 \angle -36.87° \times 104.57 \angle 69.19°$$

$$= \underline{3.41 \times 10^3 \angle 1.14°}, \quad \therefore \quad \text{regulation} = \frac{3.41 - 3.3}{3.3} = \underline{3.3\%}$$

Now

$$\bar{V}_p = \frac{\bar{E}_R \times 415}{3.3 \times 10^3} = 428.5 \angle 1.14°$$

and

$$\bar{I}_p = \frac{\bar{I}_R \times 3.3 \times 10}{415} = 9.63 \angle -36.87°$$

To get the total phase current on the delta side we must add $\bar{I}_0$ to $\bar{I}_p$ where $\bar{I}_0 = I_0 \angle -76.82°$. Remember that $\bar{I}_0$ lags $428.5 \angle 1.14°$ by $77.96°$.

$$\text{Its magnitude} = 0.462 \times \frac{429.5}{415} = 0.478 \text{ A}$$

$$\therefore \quad \text{Total phase current} = 0.478 \angle -76.82° + 9.63 \angle -36.87°$$

$$= 10.00 \angle -38.63° = \bar{I}_{BR}$$

Amplitude of line current on delta side

$$I_L = \sqrt{3}\, I_p = \underline{17.32\ \text{A}}$$

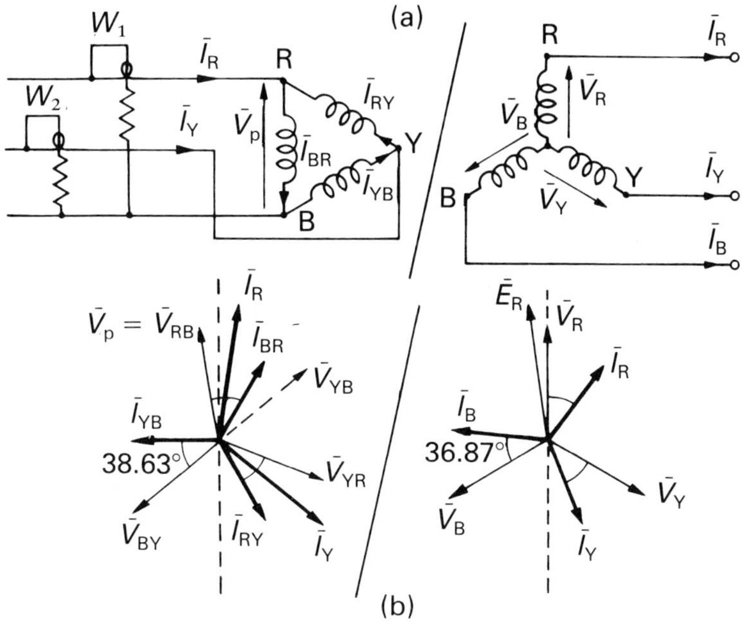

Fig. 3.14 (a) Circuit diagram showing the two wattmeter connections and (b) the phasor diagram for voltages and currents.

Fig. 3.14(a) shows the two-wattmeter connection as described in Section 1.8.5 and Fig. 3.14(b) shows the primary and secondary phasor diagrams (not to scale). Note in passing that the *line* voltages on the delta side are in phase with the *phase* voltages on the star side.

$$\bar{I}_R = \bar{I}_{BR} - \bar{I}_{RY} = 17.32 \angle -8.63^\circ \text{ A}$$

$$\bar{I}_Y = \bar{I}_{RY} - \bar{I}_{YB} = 17.32 \angle -128.63^\circ \text{ A}$$

$$\bar{V}_{RB} = 428.5 \angle 1.14^\circ \text{V}$$

$$\bar{V}_{YB} = 428.5 \angle -58.86^\circ \text{ V}$$

$$\therefore \quad W_1 = V_{RB} \cdot I_R \cdot \cos(1.14^\circ + 8.63^\circ) = Re(\bar{V}_{RB} \cdot \bar{I}_R^*)$$

$$= 428.5 \times 17.32 \cos 9.77^\circ$$

$$= \underline{7317\ \text{W}}$$

$$W_2 = V_{YB} \cdot I_Y \cos(-58.86^\circ + 128.63^\circ) = Re(\bar{V}_{YB} \cdot I^*)$$

$$= 428.5 \times 17.32 \cos(69.77°)$$

$$= \underline{2568 \text{ W}}$$

$$\therefore \quad W_1 + W_2 = \underline{9885 \text{ W}}$$

$W_1 + W_2$ should give the output power plus losses.

$$\text{Output} = 9600 \text{ W}$$

$$\text{Iron loss} = 120 \times \left(\frac{428.5}{415}\right)^2 = 128 \text{ W}$$

$$\text{Copper loss} = 150 \times \frac{1.21^2}{1.16^2} = 164 \text{ W}$$

$$\text{Hence output + losses} = \underline{9892 \text{ W}}$$

The difference of 7 W is an error of 0.07 per cent which is easily accounted for in slight rounding off errors in the figures used.

$$\therefore \quad \text{Efficiency} = 9600/9892 = 97.1\%$$

Sketching the phasor diagram is very helpful in making sure that we get the correct value for the angle between the current and voltage for a given wattmeter.

W_1 and W_2 could also have been found using the formula

$$\frac{W_1 - W_2}{W_1 + W_2} = -\frac{1}{\sqrt{3}} \tan \theta,$$

where θ is the angle between $\bar{V}_{RB}$ and $\bar{I}_{RB}$. Care must be taken in the definition of W_1 and W_2 when using this formula. See end of Section 1.8.5.

3.8 CONCLUSIONS

Subject to certain assumptions an equivalent circuit has been derived for two winding transformers. The use of the equivalent circuit in single phase and three-phase transformer problems has been demonstrated in several examples.

PROBLEMS

1 A 50 kVA, 240/4800 V, 50 Hz, single phase transformer gave the following test results:

open circuit test: 240 V, 7.5 A, 900 W.
short circuit test: 210 V, 10.5 A, 1200 W.

If the transformer supplies a load of 24 kW at 0.6 p.f. lagging and 5.0 kV, calculate the supply voltage on the low voltage side and the efficiency of the transformer with this load.

Ans. 258 V, 93.3%

2 A 500 kVA, 6.6/2.2 kV, 50 Hz, single-phase transformer gave the following test results.

Open circuit test:
LV winding: 2.2 kV, 7.5 A, 3.3 kW
HV winding: 6.6 kV, open circuit
Short circuit test:
LV winding: short circuit
HV winding: 350 V, 75.8 A, 4.2 kW

If the transformer supplies a load of 500 kVA at 0.8 p.f. lagging and 2.2 kV calculate the supply voltage on the HV side.

Ans. 6.84 kV

3 A transformer supplies a variable current to a load at constant voltage and power factor. If the iron losses in the transformer remain substantially constant show that the transformer is most efficient when the copper loss is equal to the iron losses.

A three-phase, 11 000/415 V, 50 Hz, star/delta connected transformer gave the following results for tests carried out on the high voltage winding.

	Line voltage	Line current	Three-phase power
Open circuit test	11 000 V	1.5 A	5000 W
Short circuit test	650 V	26.3 A	8000 W

At what value of line current on the low voltage side will the efficiency be a maximum? Calculate the secondary terminal voltage when supplying this current at 0.8 p.f. lagging, the primary voltage being maintained at 11 000 V.

Ans. 551 A, 399 V

4 Three 50 Hz single-phase power transformers each rated at 4 kVA, 415 V/3.3 kV are connected as shown in Fig. 3.15. The three low voltage windings are connected in delta and the three high voltage windings are connected in series with a 50 Hz voltage supply, E_s, and an ammeter I_s. With E_s switched on but set to zero, the low voltage windings are connected to a 415 V, three-phase, 50 Hz supply. Instruments on the low voltage side indicate negligible line current, total power of 120 W and line voltages of 415 V. The ammeter I_s reads zero. With the three-phase supply still connected to the low voltage windings, E_s is increased to 364 V at which value I_s reads 1.16 A and power supplied by the source E_s is 150 W. The instrument readings on the low voltage side remain substantially the same as when E_s was set to zero. Removal of the three-phase supply to the low voltage windings does not substantially alter I_s, or the power drawn from E_s.

One of the above transformers is used to supply a single-phase load of 3.2 kW at 0.8 p.f. lagging. If the load voltage is 3.3 kV, estimate the voltage applied to the low voltage winding and the transformer efficiency.

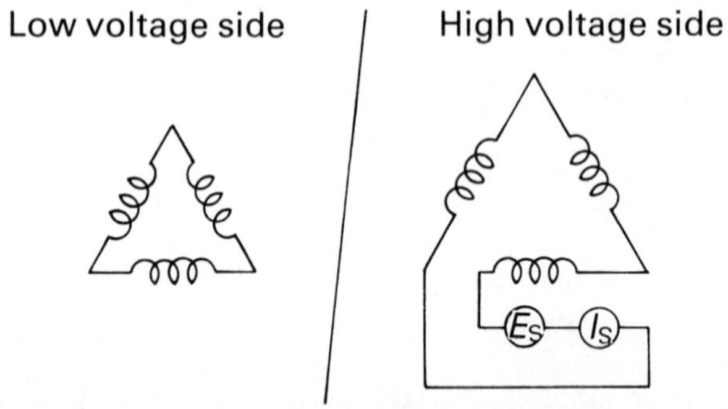

Fig. 3.15 Circuit diagram for Problem 4.

Ans. 429 V, 97%

4

Simple Machine Windings

4.1 INTRODUCTION

In order to abstract maximum power from a conventional radial air-gap electrical machine one must consider, *inter alia*, how to interconnect the current carrying conductors on each part of the machine. We shall confine our considerations to two basic winding patterns one suitable for an a.c. machine and the other suitable for a d.c. machine.

To the uninitiated, the winding pattern problem is essentially one of visualising where the conductors go on any particular member of the machine. No amount of sophisticated mathematics can help us with this problem unless we understand the basic topography of the pattern. A glance at the winding pattern of a practical machine may fill the reader with gloom when he sees the mass of lines and interconnections involved. Hopefully no such gloom awaits you in the following sections where we shall be dealing with the relatively simple task of interconnecting twelve conductors into the necessary winding type. Such a small number of conductors will suffice to illustrate the essential features of the patterns whilst at the same time keeping the diagrams simple.

4.2 CONDUCTOR E.M.F.

In Section 2.4 we saw that a sinusoidal current distribution on the inner surface of the stator of a uniform air-gap machine gave rise to a cosinusoidal flux-density distribution in that air-gap. Fig. 4.1(a) shows the stator current and air-gap flux density in such a machine and Fig. 4.1(b) shows the same air gap developed into a flat geometry with a plot of air-gap flux density against distance along the rotor surface. As explained in Section 2.4 radial flux lines appear vertical on the developed diagram. This flux is referred to as the 'field' and the stator current distribution which produces it is referred to as the 'field' winding.

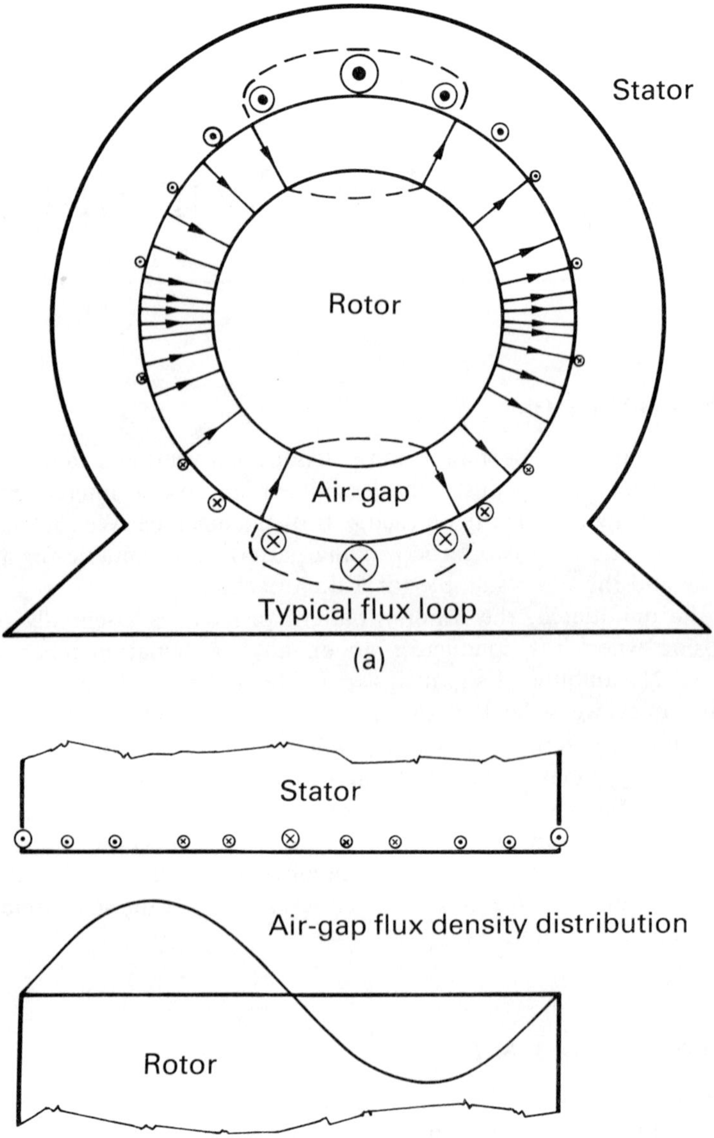

Fig. 4.1 (a) Section through uniform air-gap radial flux machine and (b) plot of radial flux density in air-gap.

Having set up a field in the air gap, we now have to consider how to place conductors on the rotor surface in order to pick up an e.m.f. The rotor is going round at angular frequency ω, say, and if we place a conductor on the rotor surface running parallel to the rotor axis then the conductor will have an e.m.f., 'e' = Blv, generated in it. B is the radial flux density at the conductor position, l is the axial length of the conductor and v is the conductor velocity. If the rotor radius is r then $v = \omega r$. B, l and v are all mutually perpendicular.

As the rotor progresses through one revolution at a steady speed the only variable in the e.m.f. equation is the flux density B. The e.m.f. will therefore follow the variation in B, i.e. it will be sinusoidal.

4.3 THE GRAMME RING WINDING

It is reasonable to suppose that in order to make best use of the available space the rotor surface should be covered with parallel conductors running in an axial direction, each generating a sinusoidal e.m.f. Because of their spatial separation on the rotor surface these conductors will, at any given instant, experience different values of B. This gives rise to a phase difference between the sinusoidal e.m.f's in different conductors. The magnitude of this phase difference will depend directly on the inter-conductor spacing.

In order to keep the diagrams simple let us not cover the rotor surface with conductors as advocated above but choose twelve evenly spaced conductors for our prototype winding. The inter-conductor spacing will be αr where $\alpha = 360^\circ/12$ and the phase difference between the e.m.f's in adjacent conductors will be α°. Fig. 4.2(a) shows the conductor spacing on the rotor surface superimposed on the field flux density distribution. Fig. 4.2(b) shows the e.m.f. variation in conductors 1 and 2 taking time $t = 0$ for the conductor positions shown in Fig. 4.2(a). Fig. 4.2(c) is the corresponding phasor diagram showing all the conductor e.m.f's. Fig. 4.2(c) immediately shows us that if we connect all of the conductors in series the resultant e.m.f. is zero and that the maximum e.m.f. we can achieve by a series connection is to take any six adjacent conductors in series. The resultant e.m.f. for the latter will have a phasor corresponding to a diameter of the envelope circle on Fig. 4.2(c).

It is important, that the conductor interconnections themselves do not contribute any additional e.m.f. The interconnections must therefore not cut the field flux and hence cannot lie on the rotor surface. The simplest and indeed, historically earliest, solution to the interconnection problem was that used in the 'Gramme Ring' winding.

The Gramme ring rotor consisted of a hollow iron cylinder attached to the shaft by means of 'spiders' at the cylinder ends. The conductors on the rotor surface were joined by interconnections passing along the inside of the iron cylinder where the magnetic field is zero. Fig. 4.3(a) shows the rotor construction and the interconnection of two adjacent conductors. Fig.

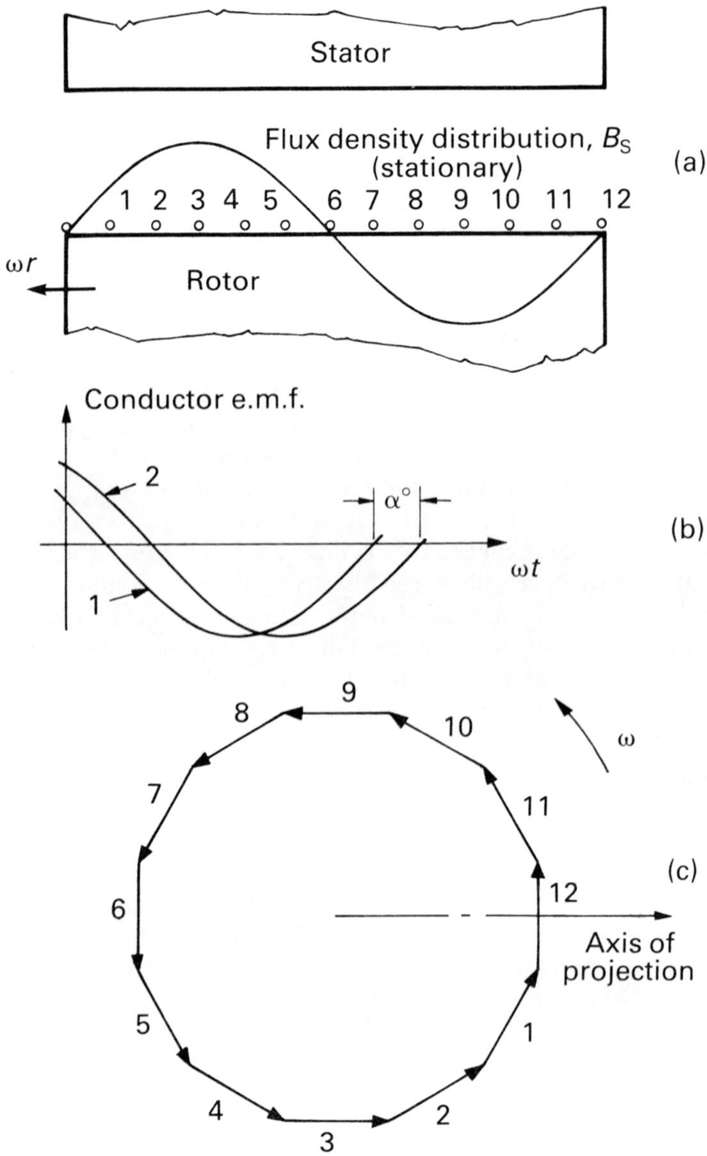

Fig. 4.2 (a) Conductor spacing and field flux density distribution, (b) e.m.f. in conductors 1 and 2, (c) phasor diagram of conductor e.m.f's.

4.3(b) shows the rotor again developed into a flat geometry and with all surface conductors joined in series. The rotor structure can be visualised as an iron toroid of rectangular cross-section carrying a closed 12 turn winding.

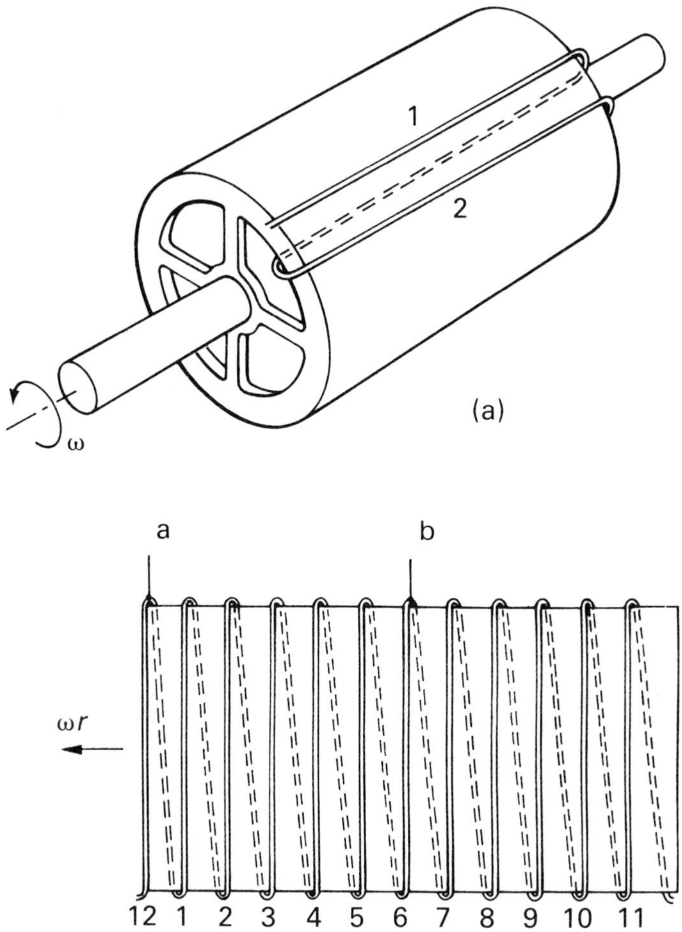

Fig. 4.3 (a) Rotor construction and (b) rotor developed into flat geometry showing closed winding.

Although the Gramme ring winding was conceived for use in d.c. machines, let us first consider its use as a single-phase a.c. winding. The largest e.m.f. is achieved by tapping off across six conductors such as at points *a* and *b* in Fig. 4.3(b). Between contacts *a* and *b* we have two sets of conductors in parallel. Either we proceed via conductors 1, 2, 3, 4, 5, 6 or via conductors 12, 11, 10, 9, 8 and 7. Note that the former path traverses the conductors from bottom to top in Fig. 4.3(b) while the latter path traverses conductors from top to bottom. In evaluating the phasor $\overline{E}_{ba}$ we need to bear in mind that the individual conductor e.m.f's must be reversed for one or other of the paths. Which path would depend on which direction we chose as reference for the conductor e.m.f's in Fig. 4.2(c). Let us assume that the path through conductors 12 → 7 is in the reference direction. Fig. 4.4 then shows the phasor $\bar{E}_{ba}$ and its constituent conductor e.m.f's.

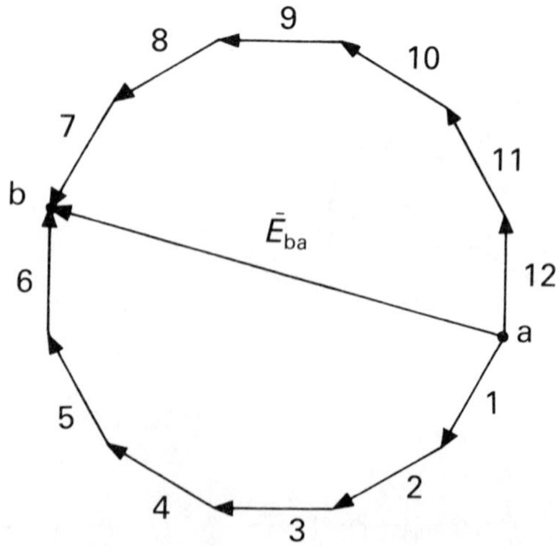

Fig. 4.4 Composition of e.m.f. $\bar{E}_{ba}$.

The connections *a* and *b* are moving at rotor velocity and in order to supply power from the winding to an external stationary load the connection to *a* and *b* must be brought out to two slip-rings on the rotor shaft and hence via rubbing contacts to the load. Fig. 4.5 shows this arrangement.

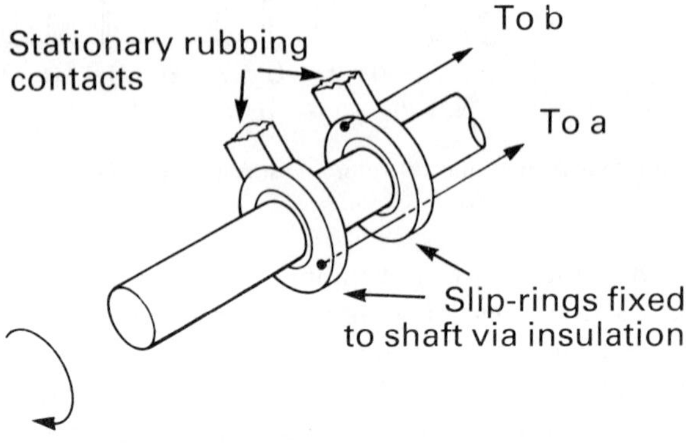

Fig. 4.5 Slip-ring connections.

The magnitude of $\bar{E}_{ba}$ is clearly *not* 6 Blv. The correct amplitude can be calculated from the geometry of Fig. 4.4 and is found to be 6 $Blvk_d$, where

k_d is called the distribution factor. In this case

$$k_d = \frac{\sin(6\alpha/2)}{6\sin(\alpha/2)} = 0.644$$

$\bar{E}_{ba}$ is a phasor of frequency $\omega/2\pi$ Hz.

A three-phase configuration can be achieved by altering the winding tap off points to R, Y and B as shown in Fig. 4.6(a).

Connections R, Y and B require three slip rings to supply an external

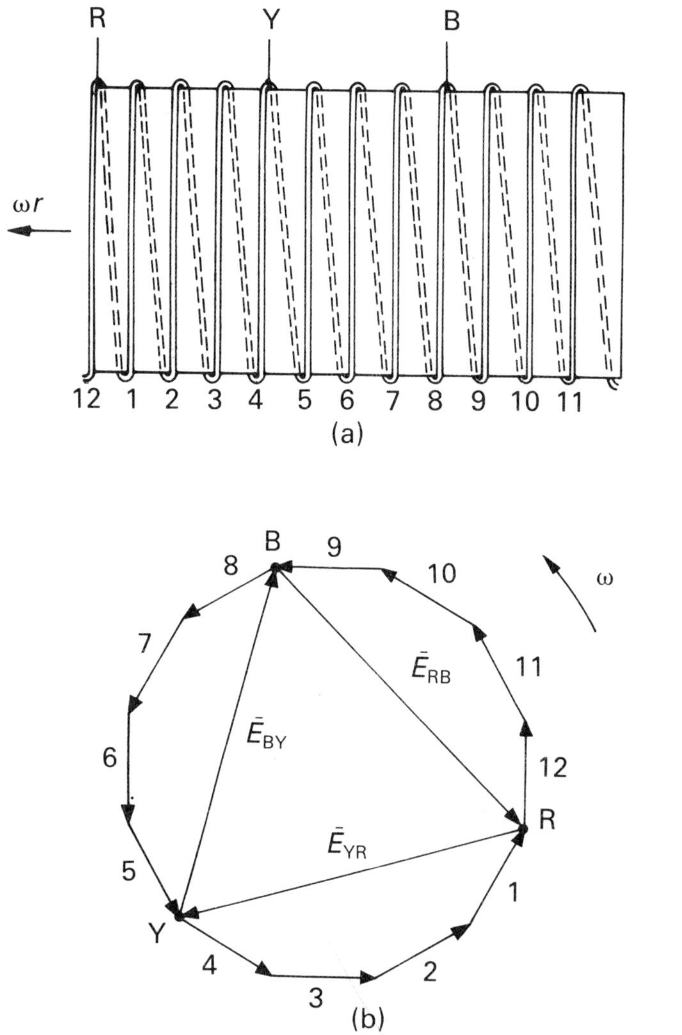

Fig. 4.6 (a) Three-phase tapping-off points and (b) phasor diagram for three-phase.

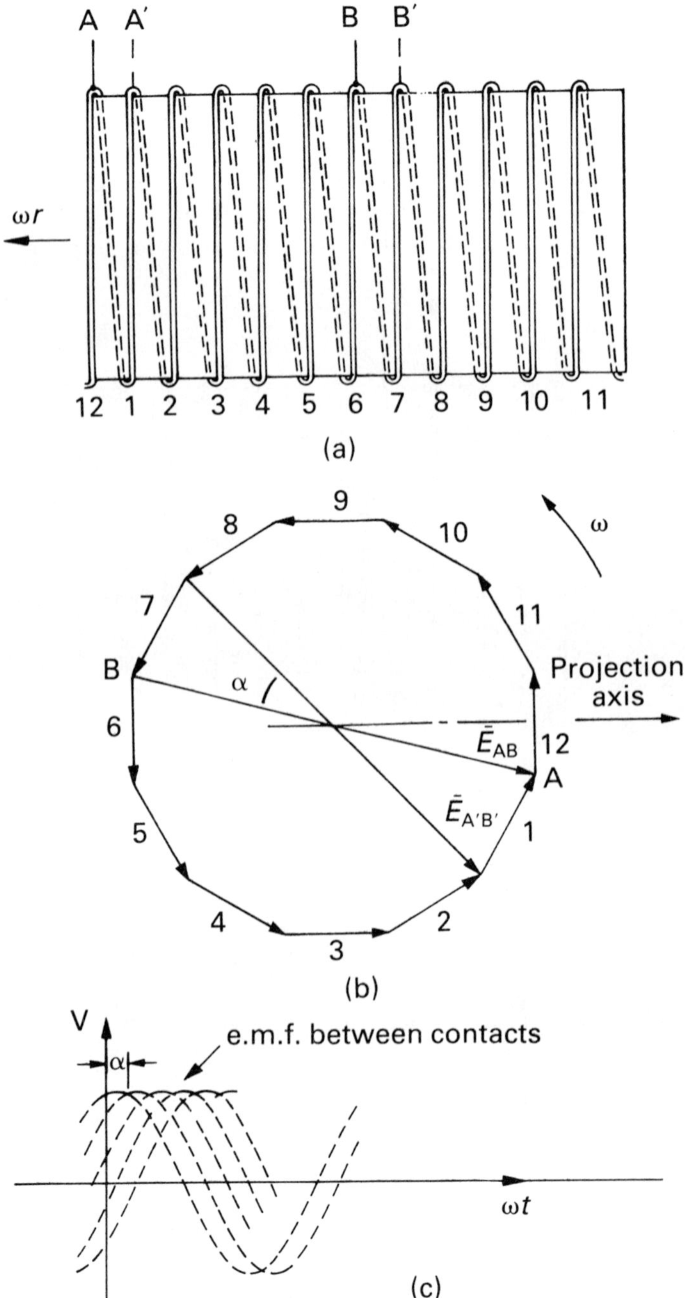

Fig. 4.7 (a) Position of tapping-off points, (b) phasor diagram, (c) e.m.f. waveform between A and B.

load. The phasor line voltages available at the slip rings are as shown in Fig. 4.6(b). The frequency of these phasors is again $\omega/2\pi$.

The one-phase and three-phase configurations are relatively simple to achieve requiring only the addition of slip rings to the shaft to facilitate connections to an external load. How might this same Gramme ring winding be connected to deliver d.c. power to an external load? It cannot be done by having tapping off points that remain fixed with respect to the rotor winding since any such connection will result in e.m.f's of frequency $\omega/2\pi$. It must therefore involve moving tapping-off points, moving, that is, with respect to the rotor winding.

Let us start from a single phase configuration with tapping off points A and B as shown in Fig. 4.7(a). A and B are not in this case connected to slip rings.

At the instant shown $\bar{E}_{AB}$ is approaching its maximum value (co-linear with the projection axis in Fig. 4.7(b)). Let A and B remain connected as in Fig. 4.7(a) for a time corresponding to a rotation of α°. At this instant contacts A and B move to positions A′ and B′ on Fig. 4.7(a). Bearing in mind that the rotor has moved forward through α° this means that A and B have been returned to their original position with respect to a stationary frame of reference. The essential difference is that the e.m.f. between the tapping off points is now $\bar{E}_{A'B'}$ as shown in Fig. 4.7(b). Fig. 4.7(c) shows the time variation of the e.m.f. between contacts A and B. It consists of the crests of successive sine waves as each α° of rotation the contacts A and B move one turn to the right on Fig. 4.7(a). Contacts A and B never move more than α° from their original position with respect to a stationary

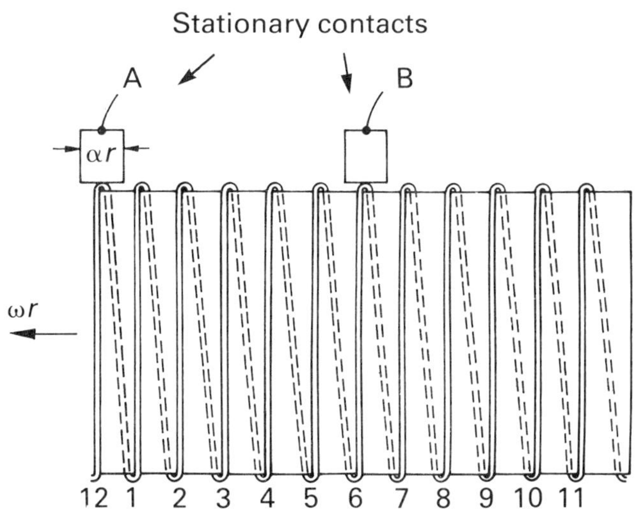

Fig. 4.8 Rotor winding showing position of stationary rubbing contacts (brushes).

reference frame. With respect to the rotor they move in a clockwise direction. The net effect of this complex process is that the e.m.f. between A and B has a large mean value (d.c.) with a small ripple superimposed.

The switching of contact points A and B seems at first sight impractical but it is achieved very simply by having two rubbing contacts each of width equal to the conductor spacing. In the early Gramme ring machines the rotor axial length was longer than the stator and the rubbing contacts bore down directly on the rotor surface where it projected from the stator. For reasons of clarity the rubbing contacts in Fig. 4.8 are shown bearing on to the edge of the winding. The contact points A and B will automatically move along the surface of the rubbing contact and then switch back to the right-hand edge as the next conductor comes into contact. Just as transfer takes place the rubbing contact will momentarily be in contact with two adjacent interconnections thereby producing a shorted turn on the rotor. What happens in this shorted turn will be considered in greater detail in Chapter 7 on d.c. machines. The rubbing contact in early machines was a 'brush' of which the 'hairs' were short flexible strands of wire. Such brushes have now been replaced by small blocks of graphite but despite this the name 'brush' is still used for the rubbing contact.

4.4 THE COMMUTATION PRINCIPLE

The process of changing the contact point from one coil to the next is called commutation. At first sight the commutation process is similar to that of rectification. A better insight into the process is gained by letting the brushes move at an angular velocity different to that of the rotor. If the brushes rotate at ω_b where $\omega_b < \omega$ then instead of contact point switching from one coil to another after $\alpha/2\pi\omega$ seconds it will now switch at $\alpha/2\pi(\omega - \omega_b)$ seconds. Fig. 4.9 shows the envelope of the voltage appearing

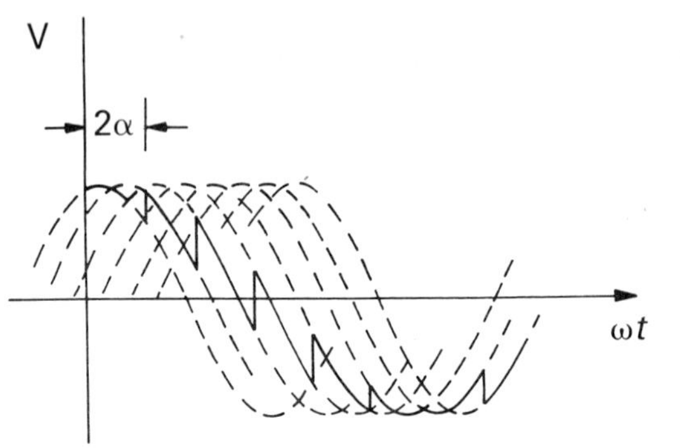

Fig. 4.9 E.m.f. waveform between brushes for $\omega_b = \omega/2$.

between the brushes for $\omega_b = \omega/2$. Apart from the ripple it is clear that we now have a sinusoidal output of half the frequency of the conductor e.m.f's. The commutation process is therefore one of *frequency* changing with the stationary brushes giving us a zero frequency output. The envelope waveform shown in Fig. 4.9 will appear again in Chapter 9 where we shall be using 'cycloconverters' to control the output frequency. Cycloconverters use thyristors to switch from one waveform to another but the principle is similar to the rotating brush situation which we have considered here.

4.5 PRACTICAL WINDINGS

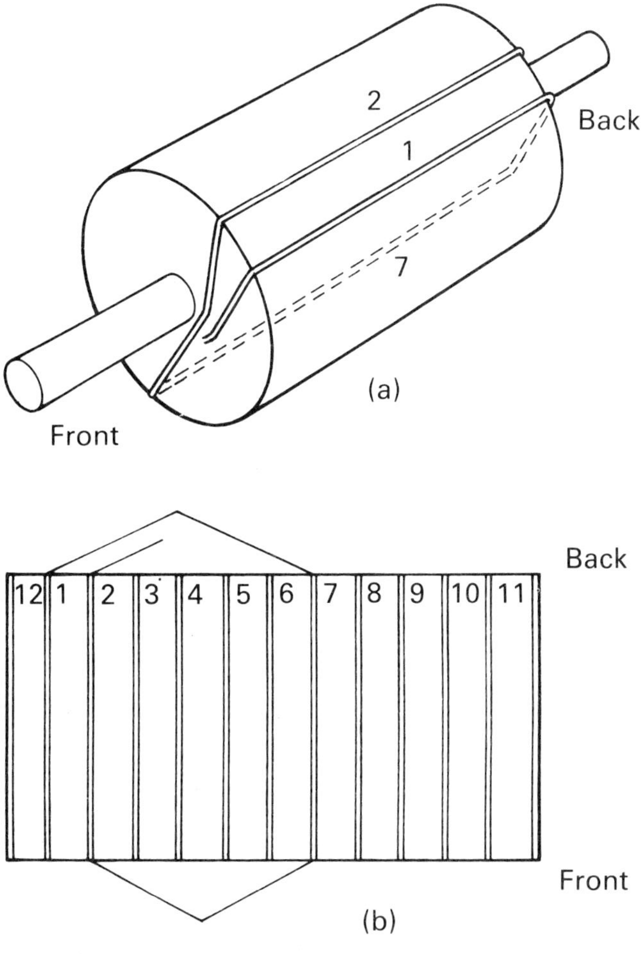

Fig. 4.10 Drum winding (a) actual shape and (b) developed winding.

The same rotor winding can therefore be used along with slip rings or brushes to give single-phase, three-phase or d.c. output voltages. All of these output connections have been illustrated by connecting a Gramme ring rotor since it is the simplest to conceive and indeed to illustrate.

The Gramme ring winding is inefficient for several reasons.

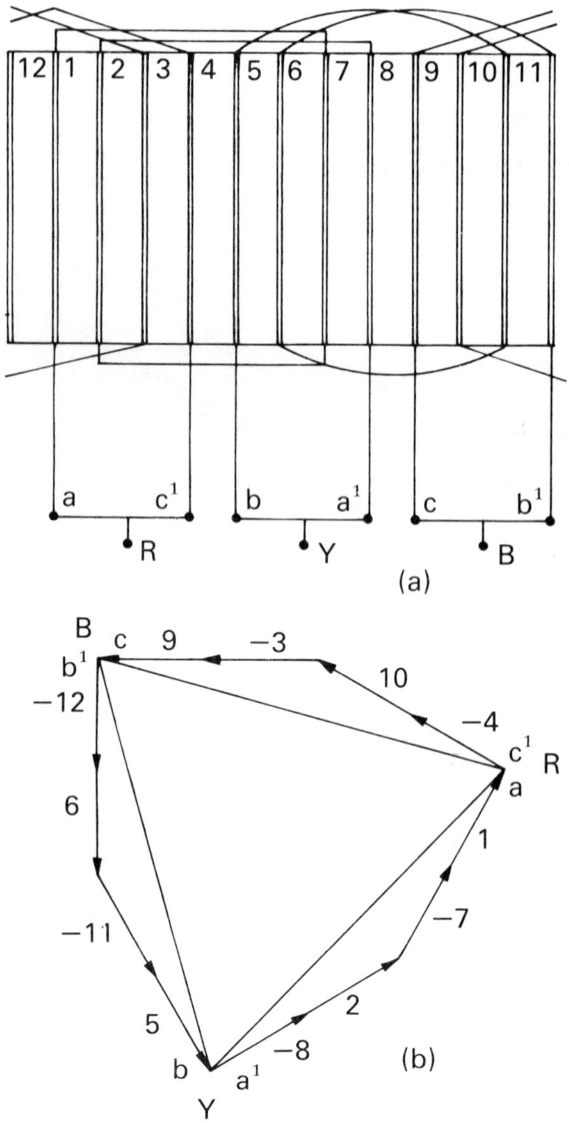

Fig. 4.11 (a) Three-phase delta-connected drum winding and (b) phasor diagram showing line voltages.

(1) If the axial length is long in comparison to the width of the iron cylinder then approximately 50 per cent of the copper in the winding is inactive, i.e. it does not lie in the magnetic field.
(2) The winding structure is not very rigid. Forces on the conductors when they carry currents have to be sustained by the conductors themselves.

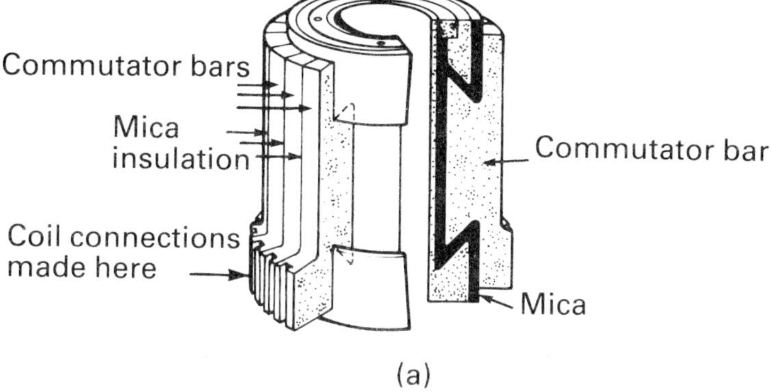

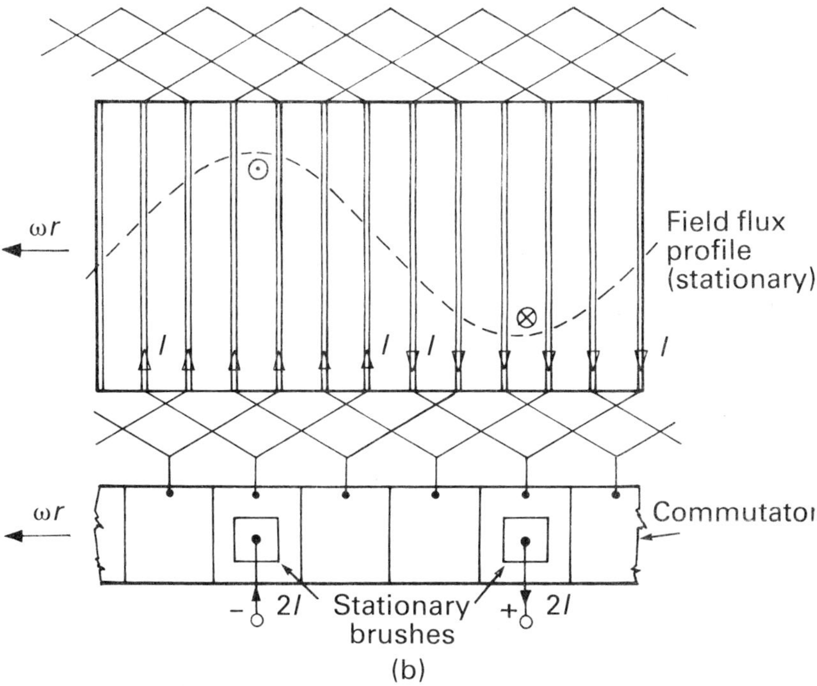

Fig. 4.12 (a) Typical commutator construction and (b) connection of coils to commutator segments.

The first point led to the development of what was called a 'drum' winding whereby instead of joining two consecutive conductors together via a connection along the back of the ring the connection is taken across to a conductor diametrically opposite at one end of the rotor and returns at the other end of the rotor to the next conductor. Fig. 4.10 illustrates this connection. The winding is as before a continuous circuit but the end connections are now shorter than the active coil sides giving a more efficient use of copper. The rotor no longer needs to be a hollow iron cylinder and can now become a solid cylinder of magnetic material.

In order to improve winding rigidity under loading conditions the conductors were then placed in slots cut in the rotor surface.

Fig. 4.11(a) shows the winding diagram for a three-phase delta connection with the points R, Y and B being brought out to slip rings. The composition of the phase e.m.f's is shown in Fig. 4.11(b). In order to help identify the phases the end connections are drawn with different shapes.

Putting the conductors in slots prevents the rubbing contacts required in the d.c. machine configuration from rubbing directly on the rotor conductors. In order to accommodate the rubbing contacts slotted windings require an additional device called a commutator which consists of a series of copper strips each insulated from the other and each connected to a coil interconnection as shown in Fig. 4.12(a) and (b).

The commutator is mounted on the shaft at one end of the machine as shown in Fig. 4.13. The brushes rub directly on the commutator segments.

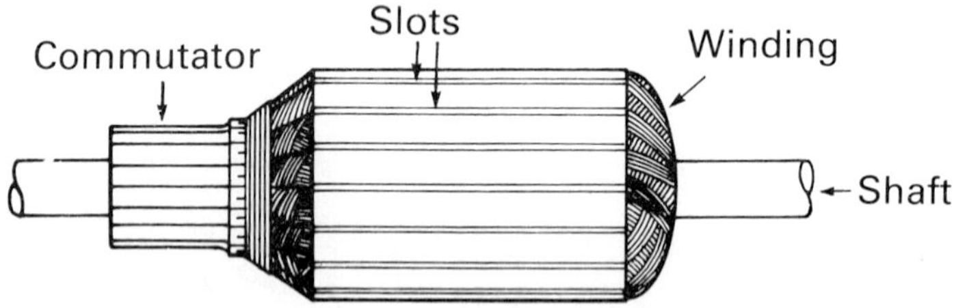

Fig. 4.13 Slotted rotor showing position of commutator.

Note that in the winding pattern of Fig. 4.12(b) the pattern progresses two coil sides at a time in comparison to the one coil side at a time in the a.c. drum windings.

4.6 THE 'ARMATURE' OF A MACHINE

The terms rotor and stator are self-evident in terms of which part of the machine they represent. The term 'armature' has no such obvious meaning but for our purposes it can be defined as that part of the machine which handles the electrical energy converted to or from mechanical energy.

So far the armature of our simple machine has been the rotor and in this respect it conforms to the normal arrangement for a d.c. machine, i.e. the 'field' winding on the stator setting up the primary flux with the 'armature' winding on the rotor. On the other hand it is normal practice for an a.c. generator to have its field winding on the rotor with d.c. currents fed to the winding via slip rings and its armature winding on the stator. In this respect therefore our simple machine has the converse arrangement to that found on practical machines. This in no way invalidates any of our deductions since they depend only on the relative motion between field and armature windings.

4.7 ARMATURE REACTION

Our field winding has been assumed to set up a sinusoidal flux density distribution in the air gap and we have so far considered only the e.m.f. generated in any particular armature winding. In order to produce electrical power from the armature winding it must obviously carry current as well as generate an e.m.f. The effects produced by these armature currents will give rise to an m.m.f. distribution in the air gap which must be superimposed on the field m.m.f. distribution. Provided the iron does not saturate, the m.m.f. distribution in the air gap will give rise to an air gap flux density distribution proportional to the m.m.f. The resultant air gap flux density can then be calculated by adding the field and armature flux density distributions. The effect on the air gap flux density produced by armature currents is called armature reaction. Let us look at some examples of armature windings considered so far.

4.7.1 Armature reaction in a d.c. winding

If we consider the generator case then the current flow will be in the same direction as the generated e.m.f. Fig. 4.14(a) shows the current and flux distribution for the brushes in the position shown in Fig. 4.12(b). The flux shown is *not* the field flux but the flux set up by armature currents acting alone, Fig. 4.14(b) is the same situation developed into a flat geometry.

Ampère's circuital law applied round a flux loop gives

$$\oint H \, dl = \text{enclosed current}$$

Neglecting the contributions to the integral from the dotted path in the rotor and stator iron and assuming H to be constant in the radial direction across the air gap then

$$\frac{1}{\mu_0} \oint B \, dl = \frac{2Bg}{\mu_0} = \text{enclosed current}$$

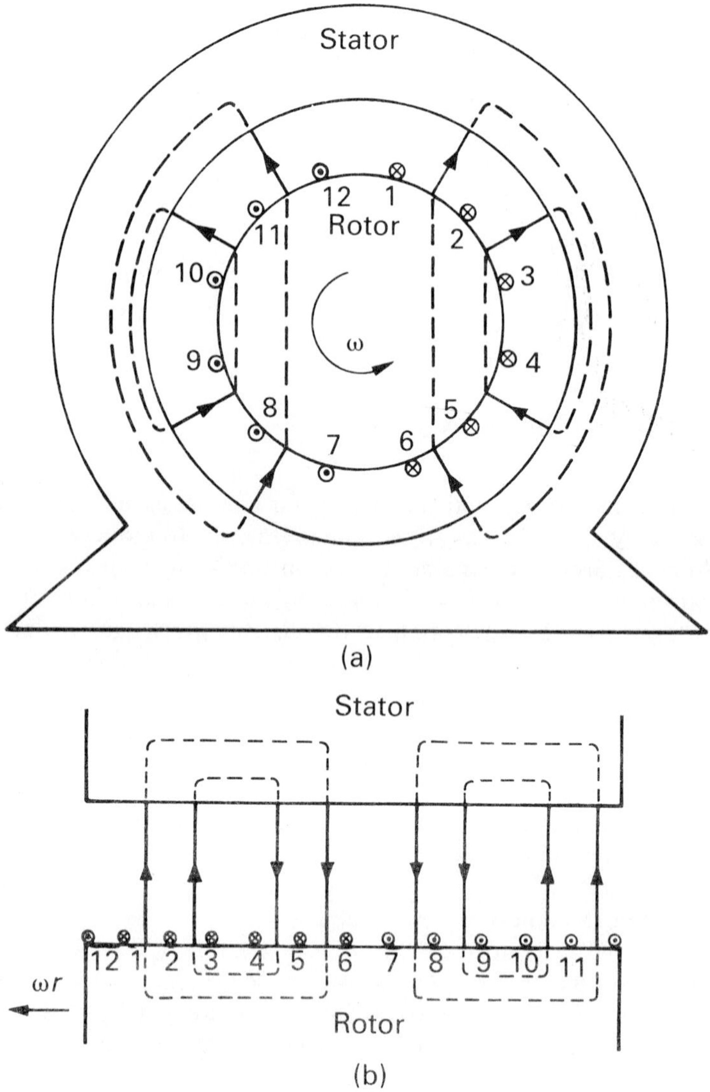

Fig. 4.14 (a) Armature currents and flux and (b) as for (a) but in flat geometry.

where g is the air gap length in the radial direction. The 2 arises from the fact that the air gap is traversed twice in completing the line integral and that B has the same magnitude at both air gap crossings of a given flux loop.

Hence for the flux loop crossing between conductors 1 and 2 in Fig. 4.14

$$B = \frac{\mu_0 4I}{2g} \text{ Tesla}$$

where I is the d.c. current flowing in conductors 2, 3, 4 and 5. By similar reasoning the flux density for the flux loop crossing between conductors 2 and 3 is given by

$$B = \frac{2I\mu_0}{2g} \text{ Tesla}$$

Fig. 4.15 shows a plot of B along the rotor surface with the field flux distribution produced by stator currents also shown.

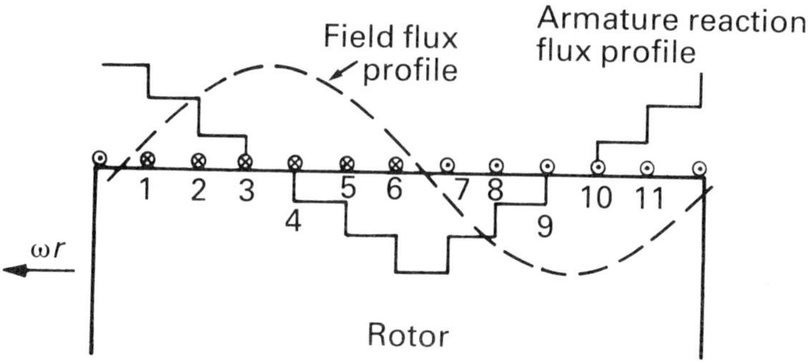

Fig. 4.15 Armature reaction flux density distribution (profile).

The torque relationship between the two flux waves is from equation (2.6)

$$T = K'B_1\ B_f \sin\theta$$

where B_1 is the amplitude of the fundamental component of the armature reaction wave and B_f is the amplitude of the field flux density distribution. θ is the angle between B_1 and B_f and can be seen to be 90° in this case.

The resultant air gap radial flux density assuming the iron circuit to be linear would be the sum of the two flux waves.

4.7.2 Armature reaction in a three-phase winding

Consider the three-phase winding configuration of Fig. 4.16(a).

The three-phase winding is placed on the stator in this case with the sinusoidally distributed field winding carrying d.c. currents on the rotor. Fig. 4.16(b) shows the flat geometry development of this configuration.

The stator currents i_a, i_b and i_c are a three-phase set and at the instant shown in Fig. 4.16(b) let us assume that i_b is at its positive maximum. The phasor diagram of Fig. 4.17 indicates that i_a and i_c will therefore be at half maximum amplitude and negative.

The contribution to the armature reaction flux density wave produced by phase b currents acting alone is as shown in Fig. 4.18. The value of B for flux loops enclosing the current is $2\hat{I}\mu_0/g$, where $\hat{I}$ is the amplitude of the phase currents.

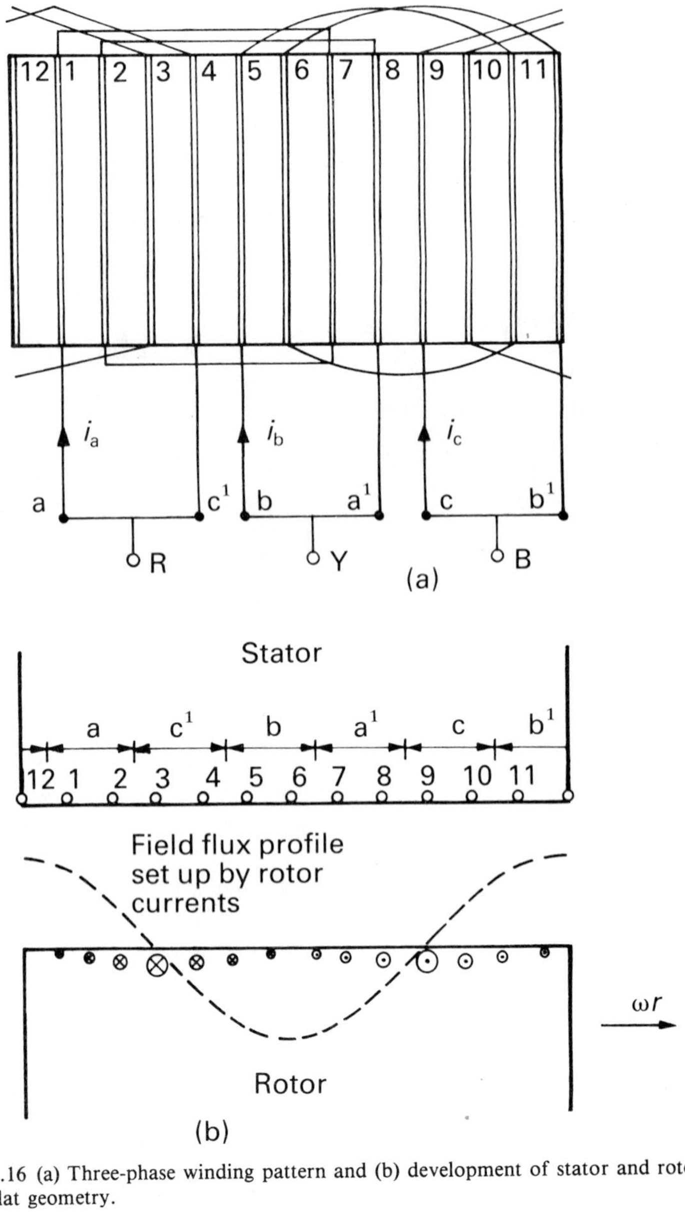

Fig. 4.16 (a) Three-phase winding pattern and (b) development of stator and rotor surfaces into flat geometry.

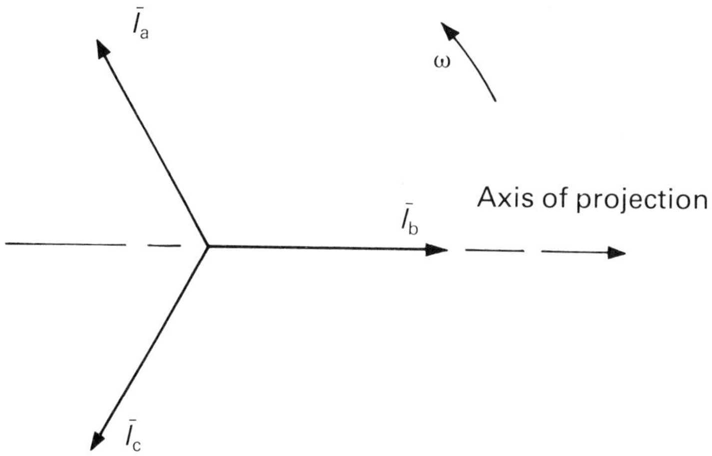

Fig. 4.17 Phasor diagram for stator currents.

By considering phase a and phase c in turn the flux density distribution due to each phase is as shown in Fig. 4.19(a) and (b). Combining the three flux density distributions again assuming linearity in the iron paths gives the resultant armature reaction flux density distribution shown in Fig. 4.19(c). The field flux density distribution is also shown in Fig. 4.19(c). Note that the peak values of the armature reaction flux density occur on the flux axis of the coil carrying maximum current, i.e. coil b.

After a time corresponding to a 120° phase delay, i_c will have reached its positive maximum with i_b and i_a being half negative maximum. The armature reaction flux density wave will have the same shape as in Fig. 4.19(c) but its peak value will occur on the flux axis of coil c. Since the rotor has also rotated through 120° in the same direction (clockwise) the relative position of the field and armature flux waves remain as in Fig. 4.19(c). After a further 120° the peak of the armature reaction flux density distribution will lie on the flux axis of coil a and the field flux distribution will have moved a corresponding amount in the same direction. The overall flux picture in the air gap is therefore that of the two flux waves rotating at the same speed in a clockwise direction one wave produced by rotor 'field' currents, the other by stator 'armature' currents.

The torque relationship between the two flux waves is as before $T = K'B_1 B_f \sin\theta$, where B_1, B_f and θ are as defined in the d.c. case. θ is no longer 90° and its value depends on the phase angle between stator e.m.f's and currents. This dependence will be considered in greater detail in Chapter 5 on synchronous machines.

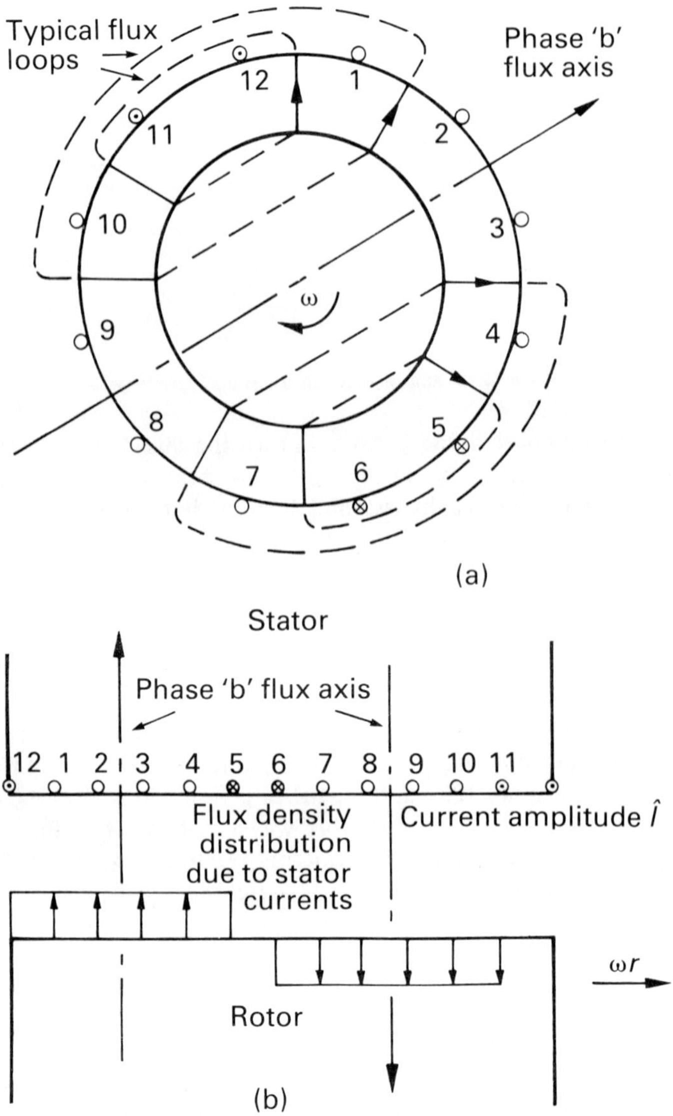

Fig. 4.18 (a) Typical flux loops for phase 'b' currents and (b) flux density distribution for phase 'b' currents.

Fig. 4.19

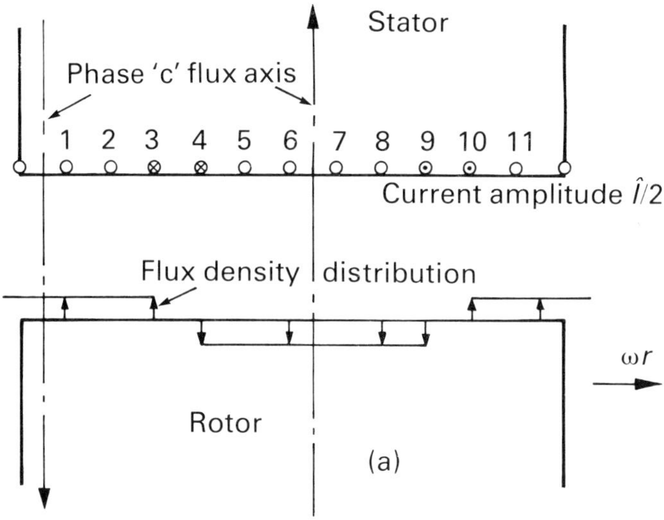
Stator
Phase 'c' flux axis
1 2 3 4 5 6 7 8 9 10 11
Current amplitude Î/2
Flux density distribution
ωr
Rotor
(a)

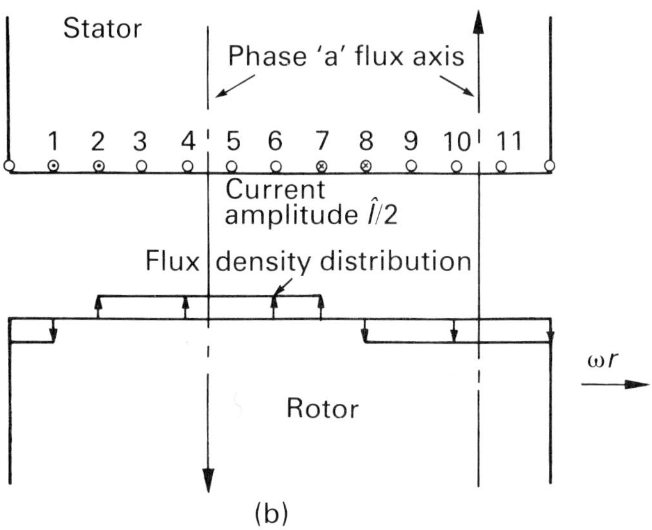
Stator
Phase 'a' flux axis
1 2 3 4 5 6 7 8 9 10 11
Current amplitude Î/2
Flux density distribution
ωr
Rotor
(b)

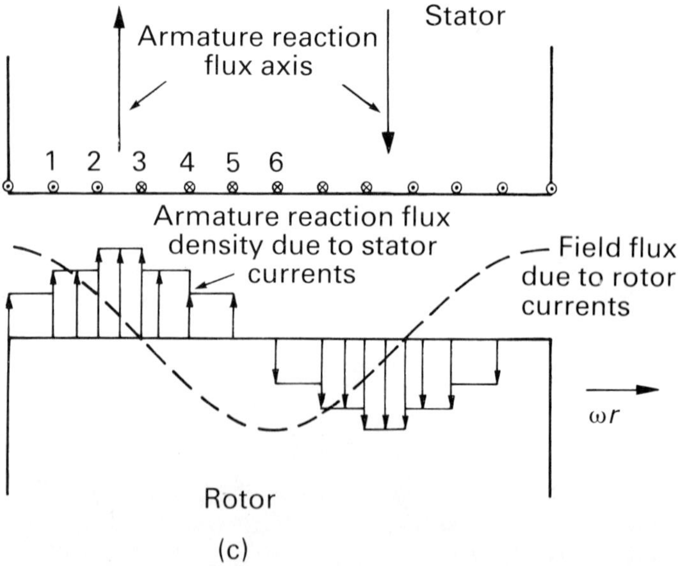

Fig. 4.19 (a) Flux density distribution for phase 'c' currents, (b) flux density distribution for phase 'a' currents and (c) resultant flux-density distribution for all three phases acting together.

4.8 CONCLUSIONS

This chapter has dealt with very simple winding patterns for d.c. and a.c. machines. Practical machine windings are beyond the scope of this text and should such information be required specialist texts such as Hindmarsh [1] or Say [2] should be consulted.

Armature reaction effects are seen to be an essential part of the torque producing mechanism and not just a parasitic side effect as many d.c. machine textbooks may lead you to believe.

PROBLEM

1 Show that the fundamental component of the flux density distribution due to stator currents in Fig. 4.19(c) is proportional to $k_d IN$, where k_d is the distribution factor for the winding shown in Fig. 4.11 and Fig. 4.19, I is the amplitude of phase currents and N is the number of turns per phase.

5

Three-Phase synchronous machines

5.1 INTRODUCTION

This chapter will deal with machines in which three-phase currents on the stator set up a rotating field in the air-gap which interacts with a field set up by d.c. currents on the rotor. The rotor must therefore rotate at the same speed as the stator derived rotating field, i.e. the rotor goes round in synchronism with the rotating field.

All the three-phase generators coupled in to electrical power grids are synchronous generators and synchronous motors are used in applications where constant speed over a range of torques is required.

The reader should be familiar with the derivation of air-gap flux densities covered in Chapter 2 and the simple three-phase winding pattern given in Chapter 4.

5.2 SYMBOLIC REPRESENTATION OF THE MACHINE

The sinusoidal flux density distributions shown in the flat geometry of Fig. 5.1(a) will be represented in the original geometry as shown in Fig. 5.1(b). This is the representation already adopted in Section 2.8.

Likewise the three-phase winding of Fig. 2.13(a) will be represented by a single coil placed centrally in the space occupied by the phase it represents as in Fig. 5.2(a). In the original round geometry this is as shown in Fig. 5.2(b).

The e.m.f. in a given phase will be at its maximum when the e.m.f. in the coil representing that phase is at its maximum. The amplitude relationship between the e.m.f. in this coil side compared to the e.m.f. in the two actual coil sides it represents can be calculated from Fig. 4.6(b) but its value need not concern us.

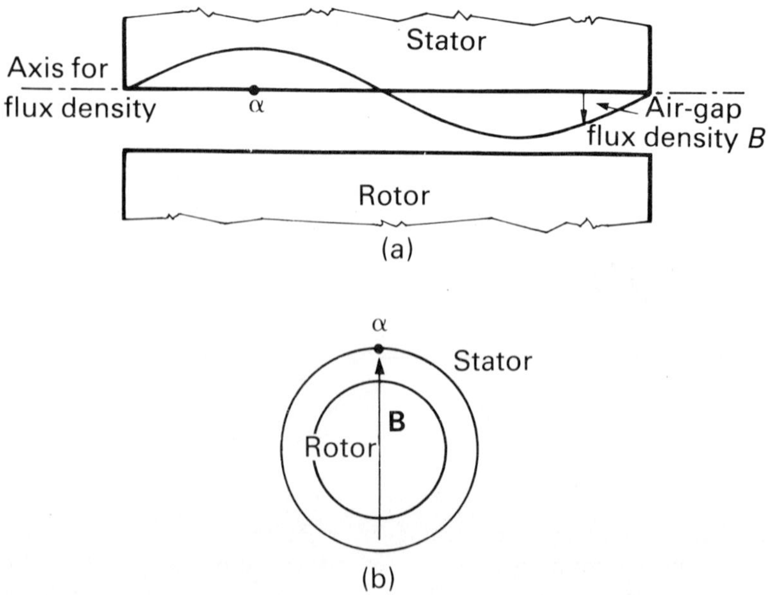

Fig. 5.1 (a) Air-gap sinusoidal flux density distribution in flat geometry and (b) vector representation of (a) in round geometry.

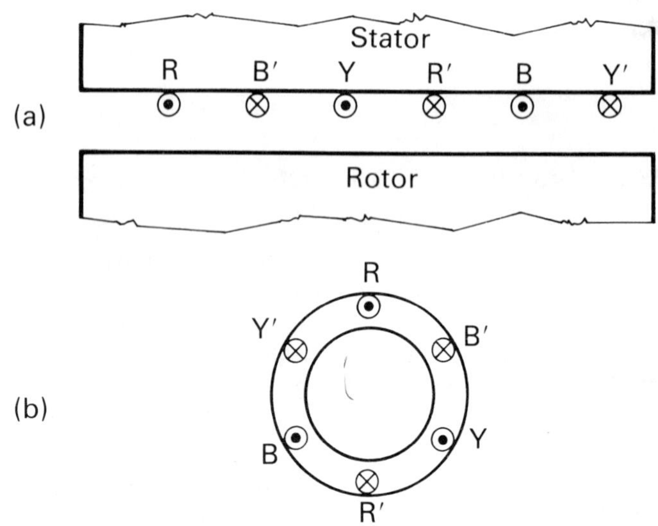

Fig. 5.2 Representation of three-phase stator winding (a) in flat geometry and (b) in round geometry.

The rotor derived flux density distribution B_r will be assumed to be sinusoidal as shown in Fig. 5.1(a). (The rotor current distribution to give such a flux wave would require to be cosinusoidal.) The rotor winding will be shown as a single coil with its sides coinciding with the maximum and minimum of the actual current distribution. Fig. 5.3(a) shows this arrangement and Fig. 5.3(b) shows the complete symbolic representation of stator and rotor currents and flux waves. The stator currents are three-phase a.c. and the rotor current is d.c.

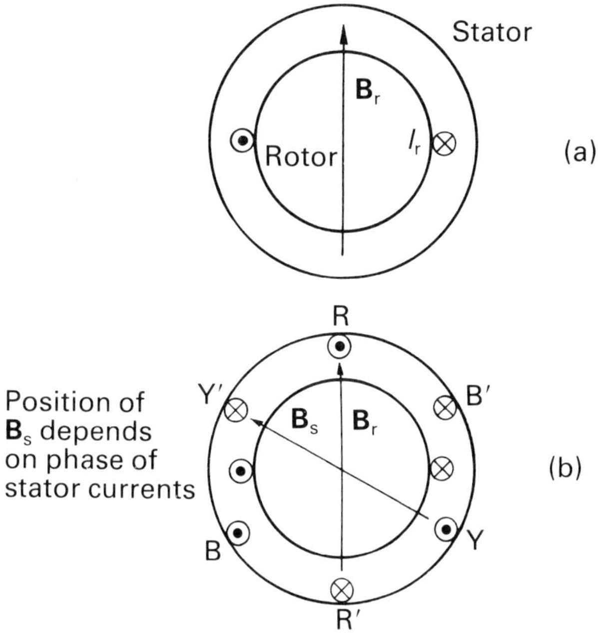

Fig. 5.3 (a) Symbolic representation of rotor currents and rotor derived air-gap flux density, (b) symbolic representation of rotor and stator currents and flux density distributions.

5.3 EQUIVALENT CIRCUIT FOR SYNCHRONOUS MACHINES

Let us for the moment ignore any losses, electrical or mechanical, in the machine. As in Chapter 2 we shall also assume that the reluctance of the flux path is constant thereby allowing us to add fluxes in the air-gap.

Fig. 5.4(a) shows the initial condition in which stator currents and therefore B_s is zero. The rotor is driven at ω rad/second and the d.c. rotor currents I_r set up B_r. At the instant shown the e.m.f. in phase R is going through its maximum value in the direction shown. This result is derived using the 'right-hand rule' for Blv remembering that v is the velocity of the

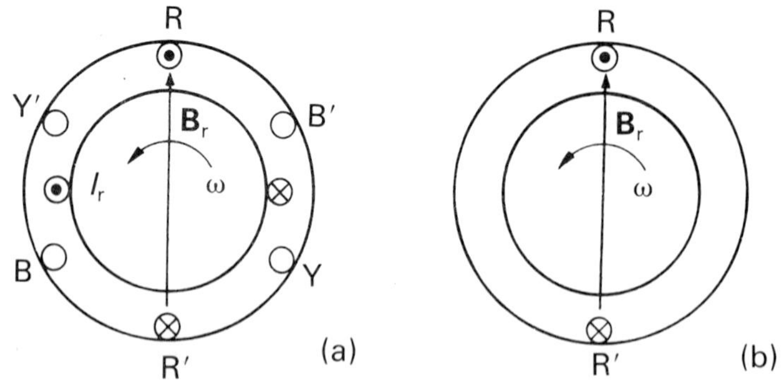

Fig. 5.4 (a) E.m.f. set up in coil R by $\mathbf{B}_r$ when stator currents are zero, (b) reduced representation for simplicity.

conductor with respect to the field. As the cosinusoidal flux distribution represented by $\mathbf{B}_r$ sweeps past the three coils it will generate a set of three-phase e.m.f's of phase sequence RBY. What happens in the R coil will happen 120° later in the B coil and 240° later in the Y coil. Our task will be considerably simplified if we make use of this three-phase symmetry and confine our deliberations to what happens in phase R. In this way we can reduce our symbolic representation to that of Fig. 5.4(b) with phases B and Y implied but not shown and the rotor effects represented by $\mathbf{B}_r$ only. The more simplification we make in our symbolic representation the more we lose track of the actual machine, but against this must be weighed the greater simplicity of the diagrams.

Let the e.m.f. generated in coil R be $\bar{E}_R$. There are no stator currents and hence no torque or power involved in driving the rotor at ω rad/s. If the torque is to be kept at zero when three-phase stator currents $\bar{I}_R$, $\bar{I}_Y$ and $\bar{I}_B$ are injected into the stator windings then the rotating flux wave B_s produced by these currents must coincide with B_r such that $B_s B_r \sin \delta$ is zero as a result of δ being zero (or π). On our simplified diagram this means that $\mathbf{B}_s$ and $\mathbf{B}_r$ are colinear as shown in Fig. 5.5(a). The generated e.m.f. $\bar{V}_R$ caused by the resultant flux wave $(\mathbf{B}_s + \mathbf{B}_r)$ will be larger than $\bar{E}_R$ and in phase with $\bar{E}_R$. The amplitude relationship is given by $E_R/V_R = B_r/(B_s + B_r)$. What must be the phase relationship between $\bar{I}_R$ and $\bar{E}_R$ to give $\mathbf{B}_s$ coincident with $\mathbf{B}_r$? This is best seen from Fig. 5.5(b) where the flux waves have advanced by quarter of a revolution to coincide with the flux-axis of coil R. From Section 2.8 we know that $\mathbf{B}_s$ coincides with the flux axis of a coil when that coil carries maximum current. $\bar{I}_R$ is therefore at its positive maximum in the direction shown in Fig. 5.5(b).

At this instant $\bar{V}_R$ is going through zero and there is therefore a 90° phase relationship between $\bar{V}_R$ and $\bar{I}_R$. This is consistent with the zero power condition arising out of $\mathbf{B}_s$ and $\mathbf{B}_r$ being coincident.

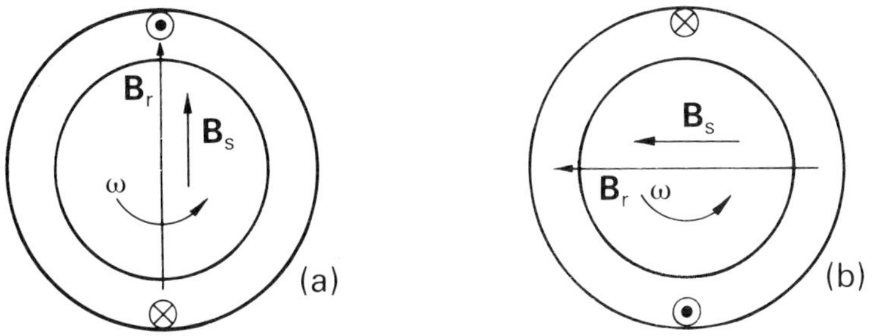

Fig. 5.5 Position of flux vectors for (a) maximum e.m.f. in coil R and (b) maximum current in coil R.

We must now consider whether the current is 90° ahead or behind the voltage. If the current is defined in the opposite direction to the e.m.f. in the coil then noting the reversal of direction between Fig. 5.5(a) and (b), the current is 90° behind the e.m.f. If on the other hand the current is defined in the same direction as the e.m.f. then the current is 270° behind (or 90° ahead of) the e.m.f. This should become clearer if we consider how we might represent the above behaviour in terms of an equivalent circuit.

With no stator current we measured a terminal voltage $\bar{E}_R$ in phase R so we may regard the coil as a voltage source $\bar{E}_R$. When stator current flows the terminal voltage alters to $\bar{V}_R$ and we may attribute the difference $\bar{V}_R - \bar{E}_R$ to a voltage difference across an internal impedance $\bar{Z}_R$ say. Fig. 5.6(a) shows the defined direction for $\bar{V}_R$, $\bar{E}_R$ and $\bar{I}_R$. Note that $\bar{I}_R$ is in the

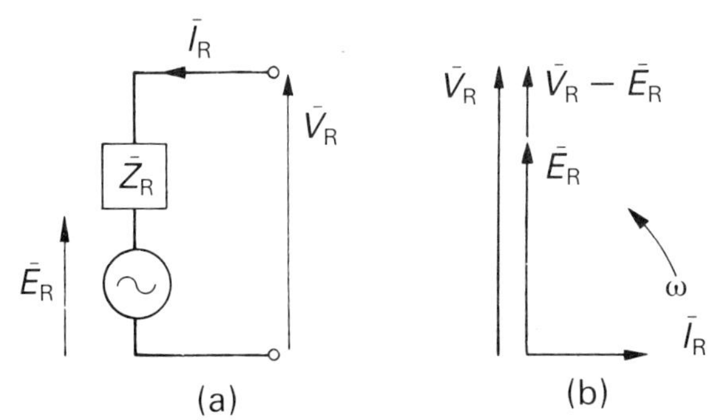

Fig. 5.6 (a) Equivalent circuit for phase R and (b) phasor diagram for flux conditions shown in Fig. 5.5.

opposite direction to $\bar{V}_R$ and $\bar{E}_R$ and is therefore in a direction consistent with the coil being considered as a power sink. As explained previously $\bar{I}_R$ so defined is 90° behind $\bar{E}_R$ and $\bar{V}_R$ and Fig. 5.6(b) shows the corresponding phasor diagram. From Fig. 5.6(a)

$$\bar{V}_R - \bar{E}_R = \bar{I}_R \bar{Z}_R$$

and since $\bar{I}_R \bar{Z}_R$ is 90° ahead of $\bar{I}_R$, $\bar{Z}_R$ must be an inductive reactance, jX_R say. Furthermore X_R is constant since the difference between $\bar{V}_R$ and $\bar{E}_R$ is directly proportional to B_s which in turn is directly proportional to $\bar{I}_R$. Fig. 5.7(a) shows the equivalent circuit per phase for the machine with $\bar{I}_R$ defined as flowing out of the terminals, i.e. consistent with a generator or power source. The corresponding phasor diagram is as shown in Fig. 5.7(b) with $\bar{I}_R$ reversed in direction compared to Fig. 5.6(b).

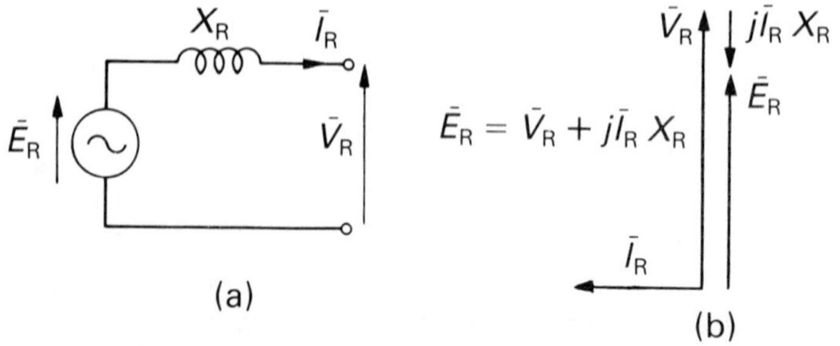

Fig. 5.7 (a) Equivalent circuit per phase with current direction as for a generator and (b) phasor diagram for flux conditions shown in Fig. 5.5.

The quantity X_R in the equivalent circuit caters for the voltage difference caused by flux set up by armature currents. This voltage difference is therefore an armature reaction effect in synchronous machines.

Fig. 5.6 and Fig. 5.7 represent exactly the same physical condition in the machine. All that has changed is our sign convention for $\bar{I}_R$. Fig. 5.6(a) is drawn with a 'motor' convention showing $\bar{I}_R$ flowing into the machine while Fig. 5.7(a) is drawn with a 'generator' convention showing $\bar{I}_R$ flowing out of the machine. There is of course no power flow since $V_R I_R \cos\theta$ is zero.

Since we have talked about current being injected into the stator windings we shall adopt the motor sign convention of Fig. 5.6(a) for the remainder of this section.

The phase relationship between $\bar{V}_R$, $\bar{E}_R$ and $\bar{I}_R$ is a direct consequence of the spatial relationship between $\mathbf{B}_r$, $\mathbf{B}_s$ and $(\mathbf{B}_r + \mathbf{B}_s)$. $\bar{V}_R$ is caused by the resultant flux wave $(\mathbf{B}_s + \mathbf{B}_r)$ sweeping past the conductors while $\bar{E}_R$ is due to the rotor derived flux wave $(\mathbf{B}_r)$ sweeping past the same conductors. In the case considered $\mathbf{B}_r$ and $\mathbf{B}_s + \mathbf{B}_r$ are colinear and hence $\bar{V}_R$ and $\bar{E}_R$ are in phase. The phase of $\bar{I}_R$ which makes $\mathbf{B}_s$ colinear with $\mathbf{B}_r$ is found to be 90° behind $\bar{E}_R$. These phase and spatial relationships can be shown side by side on the same figure as in Fig. 5.8(a) and (b). With the sign convention chosen and with the phasor diagram projection axis as shown $\bar{I}_R$ and $\mathbf{B}_s$ will be colinear but in opposite directions. $\bar{V}_R$ will be at right angles to $(\mathbf{B}_s + \mathbf{B}_r)$ and $\bar{E}_R$ at right angles to $\mathbf{B}_r$. In our linear model $|\mathbf{B}_s| \propto |\bar{I}_R|$, $|\bar{V}_R| \propto |(\mathbf{B}_r + \mathbf{B}_s)|$ and $|\bar{E}_R| \propto |\mathbf{B}_r|$.

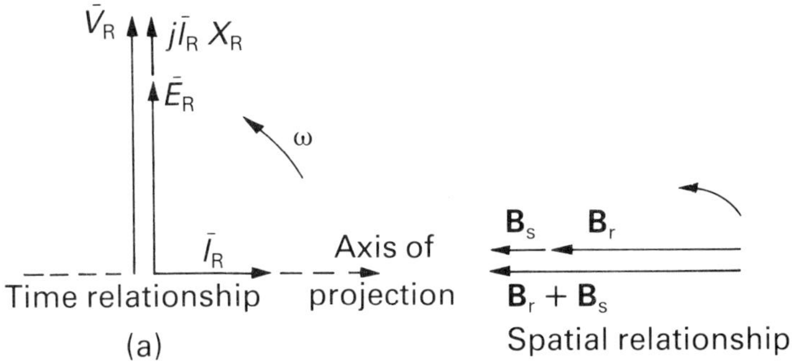

Fig. 5.8 (a) Phasor diagram and (b) flux density vectors for conditions shown in Fig. 5.5.

We must be careful not to confuse $\mathbf{B}_r$ and $\mathbf{B}_s$ with phasors. They represent cosinusoidally distributed flux waves whose position alters with time but whose amplitude remains constant.

Let us now examine what happens when we apply a brake torque to the rotor shaft while maintaining $\bar{I}_R$ and I_r as in the previous case. The positions of $\bar{I}_R$ and $\mathbf{B}_s$ will remain as in Fig. 5.8 and the amplitude of $\mathbf{B}_r$ will remain as in Fig. 5.8(b). The application of an external brake torque will momentarily decelerate the rotor and shift the spatial position of $\mathbf{B}_r$ to some angle δ behind $\mathbf{B}_s$ such that $K'B_sB_r \sin\delta$ balances the applied torque (K' is a machine constant). The dynamic process by which the rotor moves from lagging $\mathbf{B}_s$ by 0° to lagging $\mathbf{B}_s$ by δ° need not concern us. Suffice to say that a new steady state condition is established with δ° between $\mathbf{B}_r$ and $\mathbf{B}_s$.

Figure 5.9(b) shows the new position of the flux vectors, $\mathbf{B}_s$, $\mathbf{B}_r$ and the resultant flux wave vector $\mathbf{B}_r + \mathbf{B}_s$.

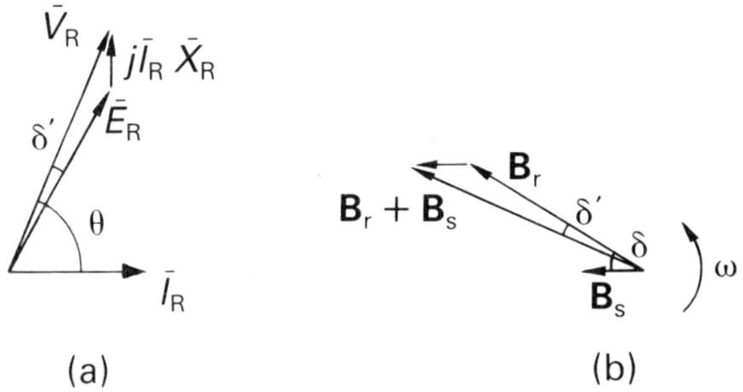

Fig. 5.9 (a) Phasor diagram and (b) flux density vectors for a loaded motor.

The phasor diagram of Fig. 5.9(a) can now be constructed bearing in mind the phase and amplitude relationships between corresponding phasors and fluxes. $(\bar{V}_R - \bar{E}_R)$ gives an identical phasor to that in Fig. 5.6(b) and since $\bar{I}_R$ is the same in both cases this verifies our equivalent circuit representation of a constant reactance in series with a voltage source $\bar{E}_R$. The amplitude and phase of $\bar{V}_R$ has changed such that

$$3V_R I_R \cos\theta = \omega T = \omega K' B_s B_r \sin\delta \qquad \text{(see equation (2.6))}$$

The angle δ on Fig. 5.9(b) is called the 'load' angle while the angle δ' on Fig. 5.9(a) and (b) is called the 'electrical load' angle.

The constraint of constant $\bar{I}_R$ imposed on the machine in the foregoing analysis is unlikely to occur in practice. Synchronous machines are invariably operated at constant terminal voltage. This results from their connection to a constant voltage power grid system. Let us therefore look at the situation whereby the same torque/power is developed as in Fig. 5.9 but this time with $\bar{V}_R$ and I_r held constant. Since $\bar{V}_R$ is constant then $(\mathbf{B}_r + \mathbf{B}_s)$ is constant. The magnitude of $\mathbf{B}_r$ will be constant as a result of constant I_r and this implies that the magnitude of $\bar{E}_R$ will be constant. Fig. 5.10(b) shows the flux distribution vectors which satisfy the above con-

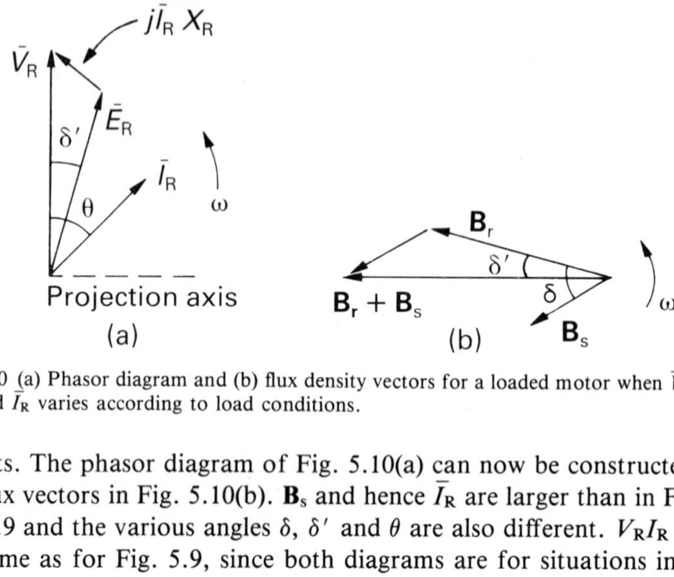

Fig. 5.10 (a) Phasor diagram and (b) flux density vectors for a loaded motor when $\bar{V}$ is constant and $\bar{I}_R$ varies according to load conditions.

straints. The phasor diagram of Fig. 5.10(a) can now be constructed from the flux vectors in Fig. 5.10(b). $\mathbf{B}_s$ and hence $\bar{I}_R$ are larger than in Figs. 5.8 and 5.9 and the various angles δ, δ' and θ are also different. $V_R I_R \cos\theta$ is the same as for Fig. 5.9, since both diagrams are for situations involving the same torque/power.

The flux distribution vector diagram has been a vital link in establishing the validity of the simple equivalent circuit of Fig. 5.6(a). Having established this equivalent circuit we can now dispense with the flux distribution vector diagram and concentrate solely on the phasor diagram arising from the interrelationship in the equivalent circuit. If we consider the rotor shaft as the output 'terminal' then all of the problems of steady state operation can be described in terms of variables measurable at the machine terminals. The physics of the conversion process need not concern us further.

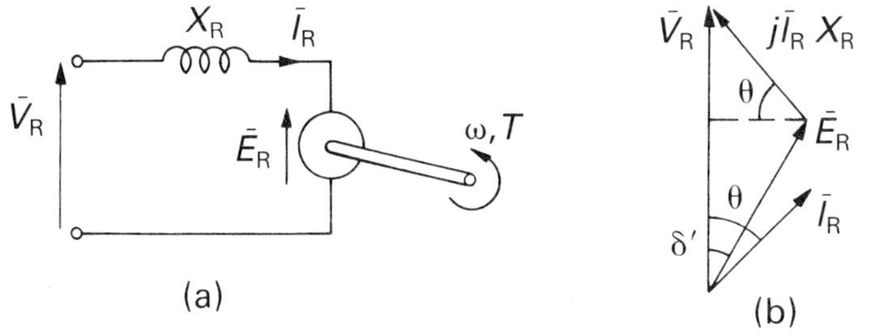

Fig. 5.11 (a) Motor equivalent circuit showing output torque and (b) corresponding phasor diagram.

Fig. 5.11 shows the motor equivalent circuit with the output 'terminal' producing a torque T.

The d.c. rotor current I_r does not appear directly on Fig. 5.11. The magnitude of $\bar{E}_R$, however, is directly proportional to I_r. I_r is most often referred to as the 'field' or 'excitation' current, and E_R is called the excitation voltage.

5.4 MACHINE TORQUE

Since there are no losses in our ideal machine the output shaft power and input electrical power are equal. Therefore

$$\omega T = 3V_R I_R \cos\theta \tag{5.1}$$

Hence

$$T = \frac{3V_R I_R \cos\theta}{\omega} \tag{5.2}$$

It is also useful to have an expression for T in terms of the electrical load angle δ' in Fig. 5.11(b). From consideration of the dotted horizontal line in Fig. 5.11(b) we can show that

$$E_R \sin\delta' = I_R X_R \cos\theta$$

and hence

$$I_R \cos\theta = \frac{E_R \sin\delta'}{X_R} \tag{5.3}$$

Substituting (5.3) into (5.2) gives

$$T = \frac{3V_R E_R \sin\delta'}{\omega X_R} \tag{5.4}$$

Equation (5.4) is useful in examining the maximum available torque for a machine for which V_R and E_R are fixed in magnitude. T_{max} will occur when $\delta' = \pi/2$. If the mechanical shaft torque is increased beyond this value by an external shaft load then δ' will increase but the 'electrical' torque given by equation (5.4) will decrease. This will leave a greater net torque (T mechanical $-$ T electrical) available to decelerate the rotor and the machine will not revert to a steady state speed of ω. The machine is said to lose synchronism or 'fall or pull out of step'. The maximum torque is also often referred to as the 'pull-out' torque since it describes the torque limit prior to the machine pulling out of step.

In our previous fundamental expression for torque in terms of the stator and rotor flux waves, equation (2.6),

$$T = K'B_r B_s \sin \delta$$

we must now remember that although B_r is fixed in magnitude (E_R constant) B_s is not of fixed magnitude. The maximum torque is therefore not when $\delta = \pi/2$ but when $B_s \sin \delta$ is a maximum. The latter condition is satisfied when $\delta' = \pi/2$.

5.5 SYNCHRONOUS IMPEDANCE

Just as in the case of the transformer we must now modify the ideal lossless equivalent circuit to take account of the imperfections of an actual machine.

The rotor circuit carries a d.c. current fed in through slip rings on the shaft. Rotor winding inductance does not therefore enter into the steady state equivalent circuit and the only rotor circuit quantity that need concern us is I_r and its relationship to E_R. Let us continue to assume that the iron circuit reluctance is linear and hence E_R is linearly proportional to I_r. Remember that $\bar{E}_R$ is the open circuit red phase voltage set up by B_r acting alone and that I_r is the current flowing in the rotor winding which sets up the B_r distribution.

The stator winding carries a.c. currents which set up the B_s distribution but not all of the flux set up by these windings will cross the air gap and return through the rotor iron. Such 'leakage' flux does not contribute to B_s and is not involved in the interaction with B_r. Because it links the stator winding it will give rise to a voltage drop in that winding since it is an a.c. flux. We must therefore introduce a small leakage reactance in series with the reactance X_R to cater for this voltage change. The stator winding resistance will also give rise to a voltage change and this too must be included in the series path. Incorporating these effects now gives the equivalent circuit shown in Fig. 5.12(a). The two reactances are combined to give the 'syn-

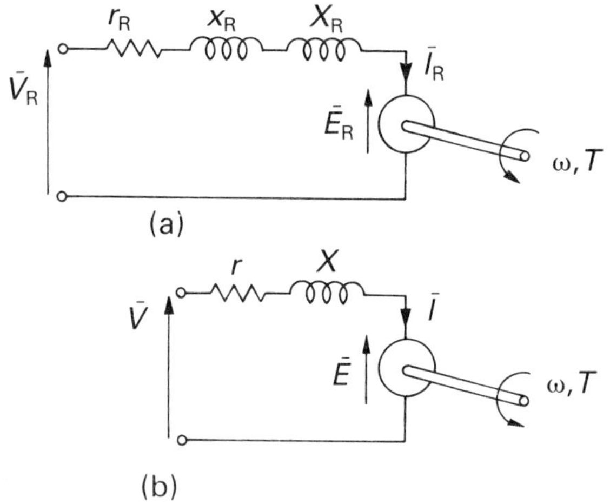

Fig. 5.12 (a) Equivalent circuit per phase showing winding resistance and leakage reactance and (b) final equivalent circuit per phase showing synchronous impedance.

chronous' reactance X per phase and $\bar{Z} = r + jX$ is called the 'synchronous' impedance per phase. Fig. 5.12(b) shows the final equivalent circuit per phase. The resistance r is typically a few per cent of X and can be neglected in calculations not concerned with stator losses. The leakage reactance x_R is also only a few per cent of X_R and we need never be concerned with the division of X into its constituent parts.

In comparison to the transformer equivalent circuit there is no element to represent iron losses. There is an a.c. flux in the stator iron and there must therefore be iron losses. When the stator current is zero and the only flux present is therefore B_r the iron losses caused by B_r sweeping past the stator iron manifest themselves as a torque resisting the rotation of the source of B_r. These losses therefore come under the heading of rotation losses and do not appear in the stator equivalent circuit. Such torque losses should be subtracted from the gross torque to give the available output torque. In a well designed machine this source of loss should be only a few per cent of the rated output. When the stator carries current there will be a small additional iron loss caused by B_s. This loss could be shown across the stator input terminals as for the transformer but it is sufficiently small to be neglected except in detailed design studies.

There will be further rotational losses such as windage and friction to add to the iron losses mentioned in the previous paragraph. Such losses are dealt with by making appropriate adjustments to the gross torque in order to arrive at the available output torque.

Fig. 5.13(a) shows the phasor diagram for a motor operating at lagging power factor and taking into account the synchronous impedance of Fig.

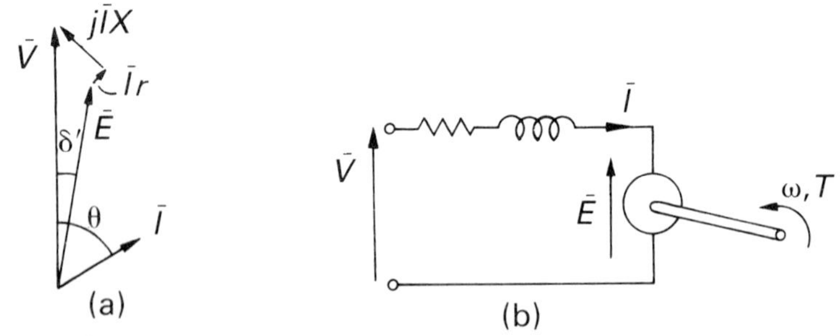

Fig. 5.13 (a) Phasor diagram and (b) equivalent circuit per phase for a motor operating at a lagging power factor.

5.13(b). The electrical input power is $3VI\cos\theta$ and the power available for conversion to mechanical work is $(3VI\cos\theta - 3I^2r)$. The quantity in brackets is also given by $3EI\cos(\theta - \delta')$. The gross torque T is given by

$$T = \frac{3EI\cos(\theta - \delta')}{\omega} \tag{5.5}$$

and the output shaft torque would be

$$T_0 = T - T_L \tag{5.6}$$

where T_L accounts for the rotational losses.

The major feature, however, of the synchronous machine equivalent circuit is the synchronous reactance X and all the significant modes of operation of synchronous machines in the steady state can be adequately dealt with by assuming a simple equivalent circuit of an ideal voltage source, $\bar{E}$, the magnitude of which depends on the level of field or excitation current I_r, and a series reactance X.

5.6 MODES OF OPERATION

Our analysis so far has assumed a 'motor' sign convention, and we have referred to output torque and input electrical power as would be the case with a synchronous motor. The synchronous machine can operate either as a motor or a generator. Once it is running at synchronous speed, ω, and connected to a fixed voltage electrical power supply its mode of operation will be determined by whatever torque is applied to the machine shaft. An applied torque resisting the shaft rotation will cause the rotor flux wave B_r to fall behind the stator flux wave B_s thereby causing the machine to operate as a motor. The rotor is being pulled round by the stator flux wave. When the applied torque is in the direction of rotation this will move the rotor flux wave B_r ahead of the stator flux wave and the machine operates as a generator. The concept of the stator flux wave being pulled round by the rotor flux wave is not particularly useful. It is better to think in terms of the stator current, $\bar{I}$, required to give $\mathbf{B}_s$ in the correct position when $\mathbf{B}_r$ is fixed in magnitude and is ahead of $\mathbf{B}_s$ and $(\mathbf{B}_s + \mathbf{B}_r)$ is fixed. With $\bar{I}$ defined using a generator convention, i.e. flowing out of the machine terminals (see Fig. 5.14(a)), then $\bar{I}$ will be in a position on the phasor diagram such that $VI \cos \theta$ is positive. The flux density distribution vectors are shown in Fig.

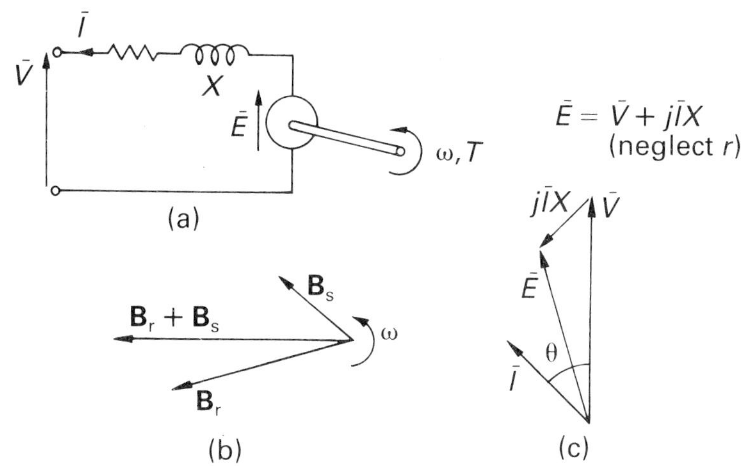

Fig. 5.14 (a) Equivalent circuit per phase, (b) flux density distribution vectors and (c) phasor diagram, for a generator operating at a leading power factor.

5.14(b) showing $\mathbf{B}_r$ ahead of $\mathbf{B}_s$ and Fig. 5.14(c) shows the corresponding phasor diagram. Note that $\bar{I}$ and $\mathbf{B}_s$ have the *same* direction since we are using a generator sign convention for $\bar{I}$.

In both the motor example of Fig. 5.13 and the generator example of Fig. 5.14, E is less than V. Such machines would be referred to as 'under-excited' machines since the open circuit voltage would be less than the supply voltage. The under-excited motor can be seen to draw power at a lagging power factor whereas the under-excited generator delivers power at a leading power factor. Both machines deliver leading vars or absorb lagging vars whichever way we care to express it. In view of the fact that the equivalent circuit contains a reactance X this might seem the obvious reason for the inductive behaviour. The error in this simple assumption can be clearly demonstrated by considering the situation whereby the power remains constant while E is increased such that $E > V$. Such machines would be described as 'over-excited'. Fig. 5.15(a) shows the motor phasor diagram and Fig. 5.15(b) the generator phasor diagram. The motor is now seen to be operating at a leading power factor while the generator is operating at a lagging power factor. The former draws leading vars while the latter delivers lagging vars. Once again, power consideration apart, both machines behave alike but in this case they look like capacitors since both draw leading vars from the supply! X is therefore not the culprit but rather it is the combination of X plus the phase and magnitude of $\bar{E}$ which determines the phase of the supply current.

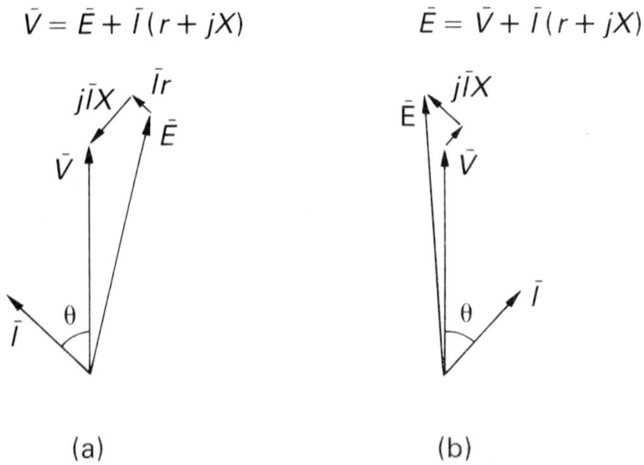

Fig. 5.15 Phasor diagrams for (a) an over-excited motor and (b) an over-excited generator.

In general, under-excited machines draw lagging (or deliver leading) vars. Increasing the excitation will decrease the lagging vars or increase the leading vars. The supply power factor can be varied, at constant power, by the simple expedient of altering the field current I_r. This feature of synchronous machines can be extremely useful in power factor correction of industrial plants where one or more synchronous motors are installed. Even in situations where there is no application for a synchronous motor an over-excited synchronous machine may be the cheapest source of lagging vars when compared to large banks of static capacitors and their associated switchgear. In such an application the machine is referred to as a synchronous 'compensator' or as a synchronous 'condenser'.

Sceptics who find it hard to accept that a lump of rotating iron plus two sets of windings can behave like a capacitor should refer to the opposite end of the power spectrum, that of microcircuits. Here they will find that inductors are simulated by a mixture of resistors, capacitors and amplifiers in what is known as a 'gyrator' circuit. The purist can rest content that both the synchronous compensator and the gyrator have transient characteristics that differ from the simple capacitor and inductor that they simulate in the steady state.

5.6.1 Power factor correction

In the previous section we saw that the power factor of a synchronous machine could be altered by adjusting the level of excitation. Fig. 5.16(a) shows one phase of the equivalent circuit of a synchronous motor. The synchronous impedance has been approximated to the synchronous reactance X and shaft torque is not shown. Fig. 5.16(b) shows the phasor diagram for the motor operating at a lagging power factor θ_1 from a constant voltage supply of $\bar{V}$ per phase. $\bar{I}_1$ and $\bar{E}_1$ are the corresponding current and excitation e.m.f. If the excitation is now increased while the load torque and power remain constant then the locus of the end of the $\bar{I}$ phasor is the dotted line AB. This locus maintains $I\cos\theta$ constant. Since $|j\bar{I}|\cos\theta$ must likewise remain constant the locus of the end of the $\bar{E}$ phasor is the dotted line CD. $\bar{E}_2$, $\bar{I}_2$ and θ_2 are for a particular value of excitation which gives a leading power factor.

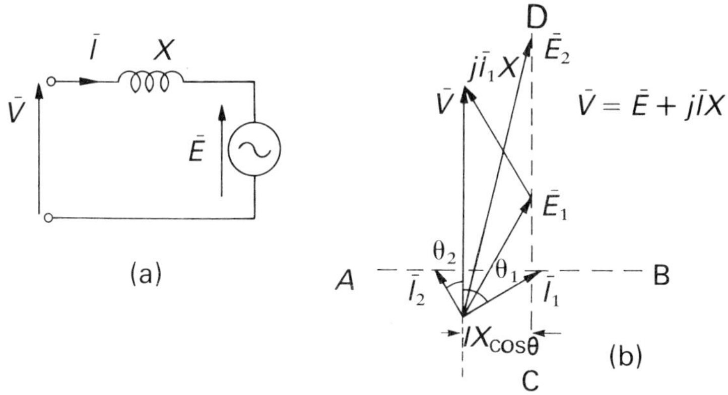

Fig. 5.16 (a) Simple equivalent circuit per phase, (b) phasor diagram for constant power at two different excitation levels ($\bar{E}$).

The leading vars drawn in the over-excited conditions can be used to offset lagging vars drawn by other parts of the installation.

Example 5.1

A 3.3 kV, 50 Hz, two-pole, star connected synchronous motor has a synchronous impedance per phase of $j15\,\Omega$. The motor is connected to a 3.3 kV, three-phase supply and its excitation is set to the rated value. What is the maximum torque the motor can develop and what is the supply power factor at this torque?

On a per-phase basis $V = 3.3/\sqrt{3}$ kV and $E = 3.3/\sqrt{3}$ kV (rated excitation). From equation (5.4)

$$T_{max} = (3 \times 3.3/\sqrt{3} \times 10^3 \times 3.3/\sqrt{3} \times 10^3)/\omega X$$

$$= \frac{3.3^2 \times 10^6}{2\pi 50.15} = \underline{2.31 \times 10^3\ \text{Nm}}$$

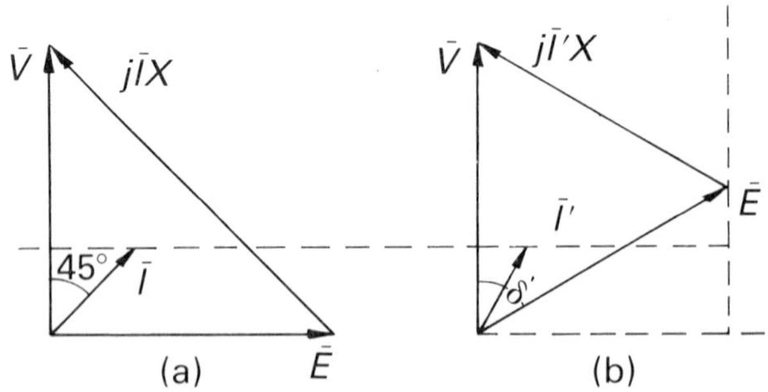

Fig. 5.17 (a) Phasor diagram for maximum torque, (b) phasor diagram for same power as (a) but with increased excitation (E).

Fig. 5.17(a) shows the phasor diagram with 90° between $\bar{V}$ and $\bar{E}$. Since $\bar{V}=\bar{E}+j\bar{I}X$, I is 45° lagging with respect to $\bar{V}$.

If the torque remains constant at this value and the excitation is increased by 20 per cent calculate the new supply power factor.

Fig. 5.17(b) shows the phasor diagram for this condition. If the voltages are drawn to scale then the construction of Fig. 5.16(b) can be used to find the direction of $\bar{I}'$ or alternately we can calculate δ' from equation (5.4) and then use $\bar{V}=\bar{E}+j\bar{I}'X$

$$T = 2.31 \times 10^3 = 3 \times \frac{3.3}{\sqrt{3}} \times 1.2 \times \frac{3.3}{\sqrt{3}} \sin \delta' \times 10^6$$

$$\therefore \quad \sin \delta' = 0.833, \ \underline{\delta' = 56^\circ}$$

$$\therefore \quad \frac{3.3}{\sqrt{3}} \times 10^3 \angle 0 = 1.2 \times \frac{3.3}{\sqrt{3}} \times 10^3 \angle -56 + j\bar{I}'X$$

Since we are only interested in the angle of $\bar{I}'$ scale I by $3.3/\sqrt{3} \times 10^3$ to get $1 \angle 0 = 1.2 \angle -56 + j\bar{I}'X'$

$$\bar{I}' = \frac{1 \angle 0 - 1.2 \angle -56}{X' \angle 90} = \frac{1.05 \angle 71^\circ}{X' \angle 90}$$

$$= \frac{1.05 \angle -19^\circ}{X'}$$

Hence the new supply power factor is $\cos 19^\circ = \underline{0.95}$

Note that $\bar{I}'$ is still lagging $\bar{V}$ despite the fact that the machine is over-excited. Increasing the excitation has none the less considerably improved the power factor. No practical machine would ever be operated at the pull-out torque since any momentary increase in load torque would lead to loss of synchronism and the motor being tripped out (isolated from the supply).

5.7 SYNCHRONOUS MACHINE TESTS

The elements of the machine equivalent circuit are determined by an open circuit and short circuit test. The machine is rotated by a separate drive motor at synchronous speed ω and the open circuit voltage is measured for a given value of field current I_r. The machine terminals are then short-circuited and I_r adjusted to give rated current in the stator windings. On the basis of a linear model the synchronous impedance is given by the ratio (open-circuit phase voltage)/(short-circuit phase current) both quantities being adjusted to correspond to the same value of I_r. Fig. 5.18(a) shows the evaluation of Z_s at the value of I_r corresponding to rated open circuit voltage. Z_s corresponds to the ratio AB/CB making due allowance for voltage and current scales. The winding resistance, r, may be measured by a d.c. test so that Z_s can then be split up into its reactive and resistive components.

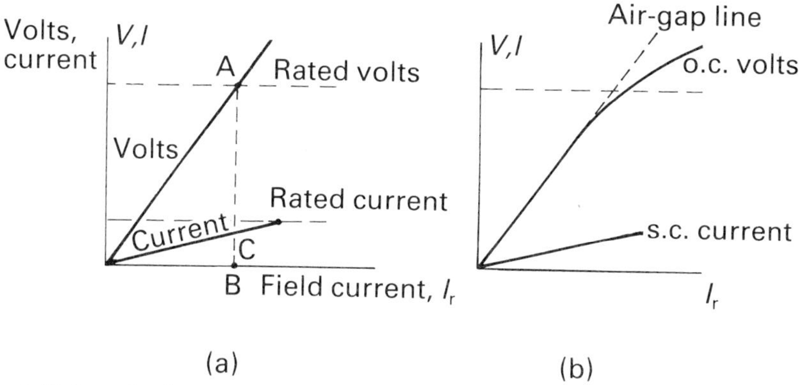

Fig. 5.18 Open circuit voltage and short circuit current plotted against-field current, I_r, for (a) linear model and (b) practical machine.

Typical values for Z_s would be such that the voltage difference across Z_s when carrying rated current would be roughly equivalent to the rated phase voltage. This is a much larger percentage voltage difference than that across the internal impedance of a transformer.

When an open circuit test is carried out on an actual machine the limitations of our linear model will immediately be evident. The graph of open circuit voltage against field current, I_r, will not be linear but will bend over as in Fig. 5.18(b). This is due to saturation effects in the iron flux return path increasing the path reluctance beyond that of the air-gap reluctance alone. While I_r is low the air-gap reluctance dominates but at high values of I_r the iron path reluctance becomes significant. The continuation of the initial slope of the open circuit curve is called the air-gap line (see Fig. 5.18(b)).

Saturation does not affect the short circuit line. The terminal voltage V is zero and hence $(\mathbf{B}_s + \mathbf{B}_r)$ is very small, sufficient only to produce a voltage to overcome the voltage difference across r and x, the leakage reactance. The short-circuit current produces a flux $\mathbf{B}_s$ which opposes $\mathbf{B}_r$ thereby keeping the net flux low.

Example 5.2

A three-phase, 10 MVA, 11 kV, star-connected synchronous machine gave the following results on test.

Open circuit line (kV) voltage	4	6	8	10	11	12	13
Field current (A)	60	90	126	171	203	252	324
Short-circuit test	Line current of 566 A at a field current of 200 A.						

Assuming r the stator winding resistance to be small calculate X from the air-gap line and short circuit test. Estimate the field current if the machine operates as a generator supplying a 10 MVA load at 0.8 p.f. lagging and 11 kV. What is the open circuit line voltage at this field?

The synchronous impedance at 200 A in the field is (from the air-gap line)

$$Z_s = \frac{13 \times 10^3}{566 \times \sqrt{3}} = 13.2\ \Omega$$

hence

$$X = \underline{13.2\ \Omega} \text{ per phase}$$

The load condition gives (per phase)

$$\bar{E} = \bar{V} + j\bar{I}X, \text{ where } |\bar{I}| \text{ is given by } \frac{10 \times 10^6}{\sqrt{3}(11 \times 10^3)} = \underline{525\ \text{A}}$$

Taking $\bar{V}$ as reference phasor

$$\bar{E} = \frac{11 \times 10^3}{\sqrt{3}} \angle 0 + j525 \angle -36.78^\circ \times 13.2$$

$$= 12.6\,\text{kV at } -19.27^\circ$$

E (line) on air-gap line $= 12.6 \times \sqrt{3} = \underline{21.8\,\text{kV}}$

The corresponding field current is <u>336 A.</u> At this value of field current the open circuit line voltage would be approximately 13.5 kV.

The numerical value for open circuit voltage although more accurate than that achieved from the linear model will not be any more than ± 10 per cent accurate. The machine is heavily over-excited and well into the saturated region.

5.8 MULTI-POLE MACHINES

For reasons of simplicity our arguments have so far been confined to two-pole machines. Such machines produce one flux density maximum (N pole) and one flux density minimum (S pole) in the B_s or B_r distribution. Fig. 5.19(a) shows the B_s and B_r distributions for a two-pole machine while Fig. 5.19(b) shows B_s and B_r for a four-pole machine. The latter has two flux maxima and two flux minima. The rotor current distribution will likewise have two maxima and two minima. The four 'poles' of the B_s distribution are produced by two distinct sets of three-phase windings each producing two 'poles'. Each three-phase winding occupies 180° of the stator surface rather than the 360° in the two-pole machines. Fig. 5.19(c) shows the stator and rotor current distributions to produce the B_s and B_r of Fig. 5.19(b).

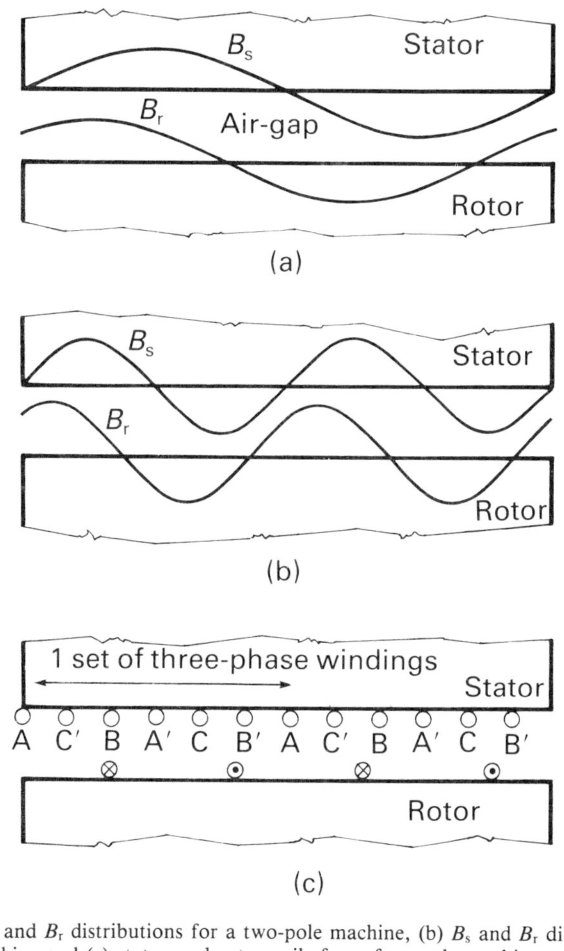

Fig. 5.19 (a) B_s and B_r distributions for a two-pole machine, (b) B_s and B_r distributions for a four-pole machine and (c) stator and rotor coils for a four-pole machine.

Like phases of the two sets of three-phase windings are probably connected in series to give a higher terminal voltage but a parallel connection would also be possible if high output current at lower voltage was required.

The shaft speed of the four-pole machine cannot remain at ω if the output frequency of the four-pole machine is to be the same as for the two-pole machine. The simplest way to relate shaft speed to frequency is to take any one pole of the B_r distribution and to consider the time, τ, it takes to sweep past one set of the stator three-phase windings. The output frequency will be $1/\tau$. For the two-pole machine a rotor pole must travel 360° to sweep past a set of stator three-phase windings. For the four-pole machines it must travel $360/2 = 180°$ to sweep past a set of three-phase windings. For a p-pole machine where p is even it must travel $(360/p/2)°$ to sweep past a three-phase set of windings. If the shaft speed is n r.p.s. then

$$\tau = (360/p/2) \div 360n = \frac{2}{np} \text{ s}$$

$$\therefore \quad f = \frac{np}{2} \text{ Hz} \tag{5.7}$$

or if

$$N = \text{r.p.m.}$$

$$f = \frac{Np}{120} \text{ Hz} \tag{5.8}$$

A six-pole machine will therefore require to run at 1000 r.p.m. to deliver power at 50 Hz or 1200 r.p.m. to deliver power at 60 Hz.

5.8.1 Salient and non-salient poles

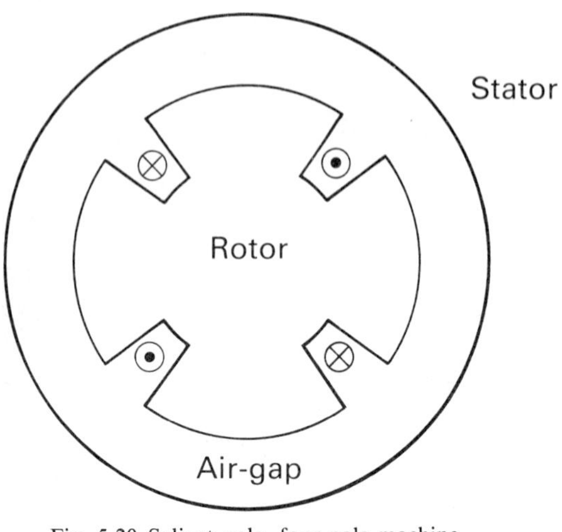

Fig. 5.20 Salient pole, four-pole machine.

Fig. 5.21 (a) A laminated pole, salient pole rotor and (b) a solid pole, salient pole rotor (Courtesy G.E.C. Large Machines Ltd.)

We have assumed a uniform air-gap in our simple linear model. Such a machine has a smooth cylindrical stator bore and a smooth cylindrical rotor. Nothing 'sticks out' into the air-gap and we would describe such a machine as a 'non-salient' pole machine. High speed turbo-alternators approximate fairly closely to a non-salient pole configuration.

The vast majority of synchronous machines have smooth cylindrical stator bores but have rotors on which the pole structure can be clearly identified from the rotor shape such as in Fig. 5.20. The poles 'stick out' into the air-gap and we would describe such a machine as a 'salient pole' machine. Fig. 5.21 shows a picture of a salient pole rotor. The air gap in salient pole machines is clearly non-uniform and the rotor current distribution is no longer sinusoidal. Careful design of the shape of the pole tip can still lead to a near sinusoidal B_r distribution. The B_s distribution is considerably affected by the rotor saliency and an accurate equivalent circuit for a salient pole machine requires the introduction of a direct axis (pole axis) reactance X_d and a quadrature axis (inter-polar axis) reactance X_q to replace the single synchronous reactance of the uniform air-gap machine. Such analysis is beyond the scope of this text and we have used only the single synchronous reactance model in our calculations. They are necessarily less accurate as a result but all the essential steady state modes of operation of salient-pole machines can be adequately dealt with by this simple model.

5.9 STARTING SYNCHRONOUS MOTORS

The B_s and B_r distributions must be rotating at the same speed in order to produce an average torque. When the rotor is stationary and three-phase stator currents of frequency f are introduced then $\mathbf{B}_s$ will have a rotational speed given by equation (5.2) while $\mathbf{B}_r$ is stationary. The inertia of the rotor is such that $\mathbf{B}_r$ cannot immediately be pulled into synchronism with $\mathbf{B}_s$ and hence no average torque exists. Synchronous motors do not therefore have any intrinsic starting torque and special precautions have to be taken to start synchronous motors.

The simplest technique when the load torque can be removed for starting purposes is to have a small auxiliary starting motor either of d.c. or induction type. Once at or near to synchronous speed the stator voltage and rotor field current can be brought to levels at which the rotor will lock into synchronism with the stator flux wave. Salient-pole motors will also develop a 'reluctance' torque which aids the locking in procedure. Once at synchronous speed rated levels of voltage and field current may be applied and the load torque brought on gradually.

The majority of synchronous motors are started making use of induction motor torques developed by currents flowing in closed loops of wire in the rotor-pole faces known as 'damper' windings. Fig. 5.21(a) shows a laminated pole rotor with damper windings embedded in the pole-face while Fig. 5.21(b) shows a rotor having solid steel poles with no damper winding. Eddy currents in the solid steel produce similar effects to the damper winding currents and the induction motor torques produced by such currents can allow starting against moderate load torques. Depending on the limitations imposed on supply currents such machines may be started at reduced or full stator voltage.

From a steady state viewpoint since $\mathbf{B}_s$ and $\mathbf{B}_r$ are not changing with respect to the rotor flux path there is no need for the rotor flux path to be laminated in order to minimise iron losses. On the other hand it may be necessary to vary $|\mathbf{B}_r|$ rapidly for control and stability purposes. In such circumstances the rotor poles must be laminated.

Where a synchronous motor is supplied from a variable frequency source such as a static inverter then the frequency of the supply can be reduced to aid starting. Inverter systems will be covered in Chapter 9 and induction motor torque in Chapter 6.

5.9.1 Synchronising synchronous generators

A generator must by definition have a prime-mover to supply the input shaft torque and this prime-mover can be used to run the generator up to synchronous speed. Before the generator terminals can be joined to the supply grid certain precautions must be taken.

(1) The generator voltage should be equal to the supply voltage.
(2) The generator phase sequence should be the same as the supply phase sequence.
(3) The generator frequency and the supply frequency should be almost identical.
(4) The voltage difference across the contacts should be very small at the instant of closing the switch.

For small machines the above process can be carried out manually and the instant at which to close the switch can be decided by connecting lights across the switch contacts as shown in Fig. 5.22(a). Fig. 5.22(b) shows the voltage across each of these lights when the generator and supply voltages are equal in magnitude but the generator frequency is slightly higher than the supply frequency. The voltage will increase to a maximum when the two phasors are in anti-phase and will go through zero when the two phasors in a given phase line up. The frequency at which the bulbs go in and out is the difference between the generator and supply frequencies. This 'beat' frequency can be adjusted to be very small by altering the speed of the generator prime-mover. This simple arrangement will also demonstrate if the phase sequence is correct. All three bulbs will go in and out together if it is correct whereas a wrong phase sequence will cause one bulb to be out when the other two are in. An additional instrument called a synchroscope can be used to indicate the precise moment when the two phasors line-up. The switch is closed at this instant provided all the other conditions have been satisfied. Once the switch is closed the machine is 'synchronised' to the supply and will run at synchronous speed thereafter unless the pull-out torque is exceeded. If the generator frequency was slightly higher than the supply frequency prior to synchronising then after synchronising the prime-mover will be trying to accelerate the rotor slightly. This will result in the machine immediately supplying power to the supply. Had the converse been true, i.e. the generator frequency was less than the supply frequency prior to synchronism, then the machine would draw power from the supply attempting to accelerate the prime-mover.

It is good practice to ascertain that the small beat frequency is obtained by the generator frequency being slightly higher prior to synchronism.

On large generators, say 100 MW or more, the closing of the synchronising switch is done automatically with various sensors checking that all the necessary conditions have been met prior to enabling the switch to be closed.

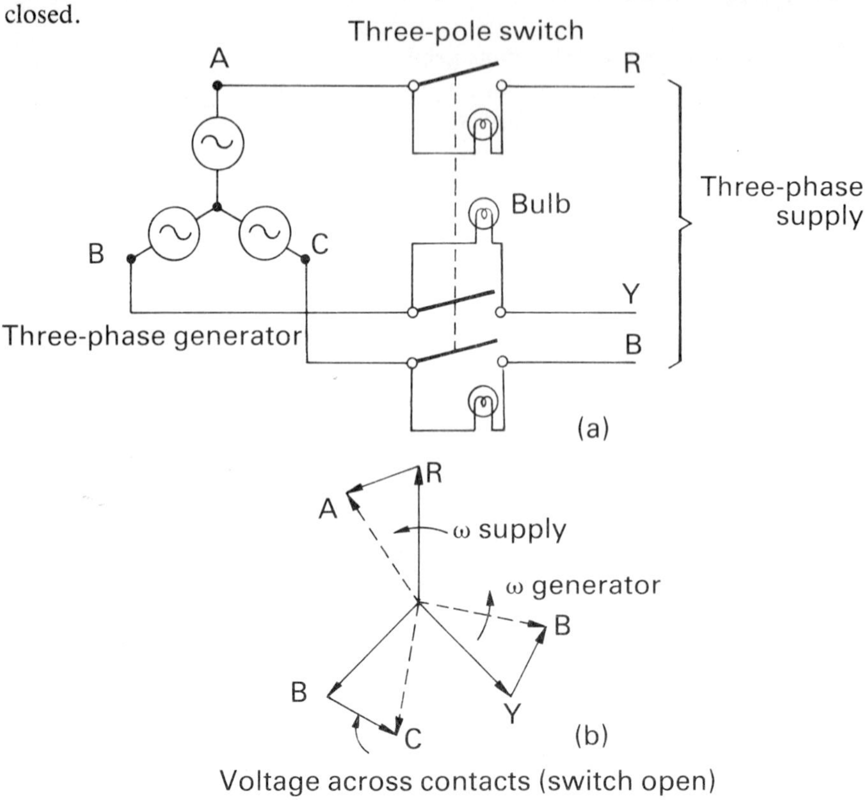

Fig. 5.22 (a) Circuit for synchronising a small generator with the three-phase supply and (b) phasor diagram showing the voltages across the open switch contacts.

5.10 EXCITATION SYSTEMS FOR SYNCHRONOUS MOTORS

During the starting and run up to speed process the rotor winding or field winding as it is most often called is either on open circuit or closed through a resistor. Once up to speed the field winding current is then introduced from an external d.c. source to allow the machine to develop the required load torque and, as we saw in Section 5.6, operate at an appropriate power factor.

There are several methods of deriving the field current. The simplest, when an appropriate d.c. supply is available, is via sliding contacts on two slip-rings on the shaft. Where no d.c. supply is available a separate 'exciter' d.c. generator is mounted on the motor shaft to produce the necessary field current. The motor drives its own exciter and control of the motor field current is usually achieved by controlling the much smaller field current of the d.c. generator. The separate d.c. exciter involves two sets of sliding contacts. One set on the d.c. machine commutator (see Section 4.5) and another set on the slip-rings of the main motor. There are now many 'brushless' excitation systems derived as shown in Fig. 5.23(a). The permanent magnet generator produces a.c. voltages on its stator winding which are then rectified and fed into a d.c. winding on the stator of the main exciter. The main exciter is an 'inside-out' a.c. generator with its a.c. windings on the rotor. Rectifiers mounted on the shaft rectify the output of the main exciter rotor windings and feed the output into the main rotor winding. The inductances of the field windings ensure adequate smoothing of the field current.

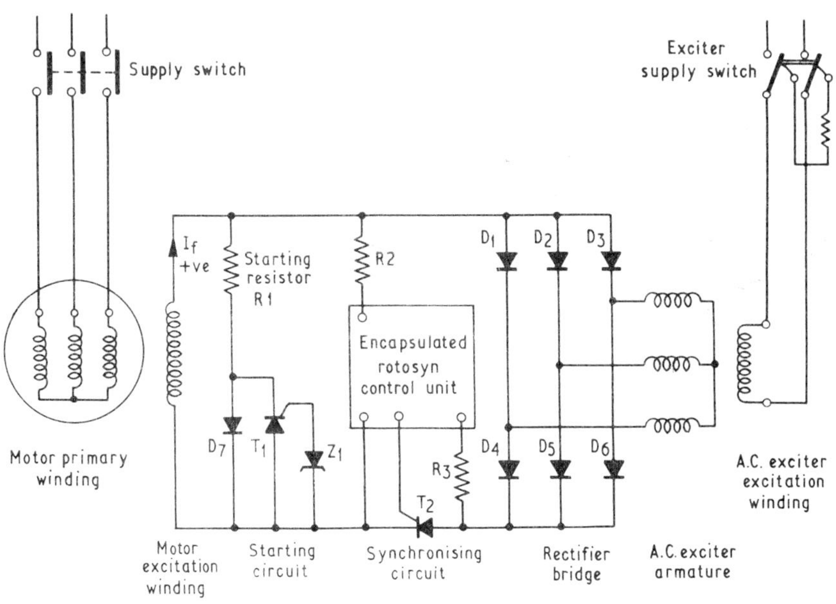

(a)

(b)

Fig. 5.23 (a) Circuit diagram, (b) photograph of a 39 000 horse-power 'brushless' synchronous motor (c) Detail of permanent magnet machine and main exciter armature. (Courtesy of G.E.C. Large Machines Ltd.)

Control of motor field current is achieved by controlling the field current of the exciter. Fig. 5.23(b) shows a photograph of a machine having the excitation system of Fig. 5.23(a). Fig. 5.23(c) shows the permanent magnet machine on the end of the main exciter armature.

5.11 APPLICATIONS

All three-phase generators with d.c. field currents come under the classification of synchronous machines. Such generators have speeds tailored to meet the characteristics of their prime movers. The turbo-alternator which forms the major power source in power networks has a high speed prime mover (steam turbine) and therefore has the minimum number of poles (2) to allow the highest shaft speed consistent with producing the desired output frequency. The speed is 3000 r.p.m. for 50 Hz or 3600 r.p.m. for 60 Hz. At high speeds the rotor radius is limited by the centrifugal forces against which the slot and winding structure must be constrained. To develop large torques and powers the active conductors on the stator must therefore be long and high speed turbo-alternators of large ratings (> 10 MW) are long and narrow. On the other hand, hydro-generators are driven by slow speed water turbines at around 100 r.p.m. This requires a machine with a large number of poles and because of the slow speed the rotor radius can be much larger than for the turbo-alternator. Because of the correspondingly large stator bore which can accommodate a large number of active conductors side by side the length of the conductors can be much less than for a turbo-alternator of the same rating. Slow speed machines tend therefore to be short axially and large radially. Fig. 5.24 shows the rotor of a turbo-alternator and of a hydro-generator.

The synchronous motor is a constant speed drive. It is invariably of salient pole construction. The efficiency of synchronous motors is higher than that of a d.c. motor or induction motor of comparable rating. It will be used therefore when either, (1) a constant speed drive without complicated control apparatus is required or, (2) a motor of high efficiency is required or, (3) power factor control is required in addition to output power. Special precautions may be required when starting synchronous motors. Synchronous motors are available in sizes up to 50 MW. Above 25 MW synchronous motors are the only types available.

Their use is widespread in process industries driving fans, blowers, conveyor belts, centrifuges, compressors, etc.

The synchronous compensator is used at strategic points in large power networks to act as a supply of leading or lagging vars. No mechanical input or output is required for such devices and apart from cooling arrangements

(a)

Fig. 5.24 (a) Long narrow rotor of a 22 MW, 3600 r.p.m., 13,8 kV, 60 Hz air-cooled turbo-alternator. (Courtesy of GEC Turbine Generators Ltd) (b) Axially short, radially large 52 pole rotor of a 36.67 MW, 11 kV, 115.4 r.p.m., 50 Hz hydro-generator (Courtesy of G.E.C. Large Machines Ltd.)

they can be completely sealed units. Reference 2 gives much more detail on the design differences between various types of synchronous machines.

5.12 CONCLUSIONS

A simple equivalent circuit has been derived for the uniform air-gap synchronous machine. Saturation effects render this circuit less accurate but the principal modes of operation of the machine can be adequately explained in terms of the simple equivalent circuit. Techniques for starting and synchronising machines have been referred to briefly, and finally a short account of machine application has been given.

PROBLEMS

1 A 22 kV, three-phase, star connected turbo-alternator with a synchronous impedance of $j1.2\ \Omega$ per phase is delivering 250 MW to the grid at 22 kV and unity power factor. With the turbine power kept constant the machine excitation is increased by 15 per cent. Assuming a linear model for the machine determine the current and power factor. At the new excitation level the turbine power is increased until the machine delivers 300 MW. Calculate the new power factor.

Ans. 6903A, 0.95(lagging); 0.985(lagging).

2 A synchronous motor has a synchronous impedance $Z \angle \alpha$ ohms per phase. Its open circuit phase voltage $\bar{E}$ lags the phase supply voltage $\bar{V}$ by δ' degrees electrical. Neglecting rotational losses show that the shaft output power is given by

$$3\left(\frac{VE}{Z}\cos(\delta' - \alpha) - \frac{E^2 \cos\alpha}{Z}\right)$$

3 A three-phase four-pole star connected synchronous motor has a resistance per phase of $1\ \Omega$ and a synchronous reactance per phase of $j10\ \Omega$. The excitation is such that the open circuit voltage would be 12 kV. When connected to an 11 kV, 50 Hz, three-phase supply calculate the maximum load the machine can take before losing synchronism. (Neglect rotational losses.) Find the corresponding current and p.f.

Ans. 11.71 MW, 887 A, 0.832 lagging.

4 A six-pole, three-phase, 50 Hz, star connected synchronous motor is supplied from 5.5 kV busbars. Its open circuit voltage is 3.3 kV per phase. The resistance and synchronous reactance are $0.5\ \Omega$ and $3.0\ \Omega$ respectively per phase. Neglecting rotational losses find the current, power factor, power and torque when the e.m.f. lags the busbar voltage by 10°, 20° and 30° (electrical degrees).

Ans. 190 A, 0.957, 1.68 MW, 16.02 kNm;
371 A, 0.995, 3.32 MW, 31.68 kNm;
552 A, 0.999, 4.80 MW, 45.87 kNm.

6

Induction Motors

6.1 INTRODUCTION

Approximately 80 per cent of all motors in use today are induction motors. They range in size from the 1/10 h.p. shaded-pole induction motors used in small domestic fans or washing machine pumps to 25 MW three-phase motors used in large industrial processes. Induction machines are rarely used in a generating mode since they require an already established a.c. supply and do not lend themselves to simple control of their generated output. We shall confine our attention therefore to induction motors starting with three-phase motors and then moving on to single-phase motors. The starting point for our explanation of the motor's operation will again be the rotating field set up by three-phase currents in a distributed stator winding. Whereas in the synchronous machine the rotor field was set up by a separately supplied d.c. current in the rotor winding, the induction motor derives its name from the fact that the rotor currents are induced by the stator field sweeping past the rotor conductors. It follows that the rotor conductors and the stator field are moving at different velocities and we shall see in our development of an equivalent circuit that this difference in speed, called the 'slip', is a very important parameter. Considerable similarity will be observed between the induction motor equivalent circuit and the transformer equivalent circuit. This is to be expected since both devices have secondary currents induced by fields set up by currents in other windings. Indeed, the induction motor with its rotor held stationary is nothing other than a somewhat inefficient transformer.

It will be assumed that the reader is familiar with the transformer equivalent circuit given in Chapter 3 and the rotating field derivation of Chapter 2.

6.2 DEVELOPMENT OF THE EQUIVALENT CIRCUIT

Both the stator and rotor carry distributed three-phase windings and this is shown diagrammatically in Fig. 6.1(a) using the same symbolic representation for each phase winding as was used in Sections 2.8 and 5.2, i.e. a single conductor shown in the centre of the section occupied by conductors of a given phase. The rotor windings are accessed via three slip-rings on the rotor shaft as shown in Fig. 6.1(b) and under normal running conditions the slip-rings are shorted together. The rotor coils are therefore closed through zero impedance and rotor coil currents are limited only by the self-impedance of their respective coil. The delta/star connection shown in Fig. 6.1(b) should not be taken to mean that all induction motors are connected this way.

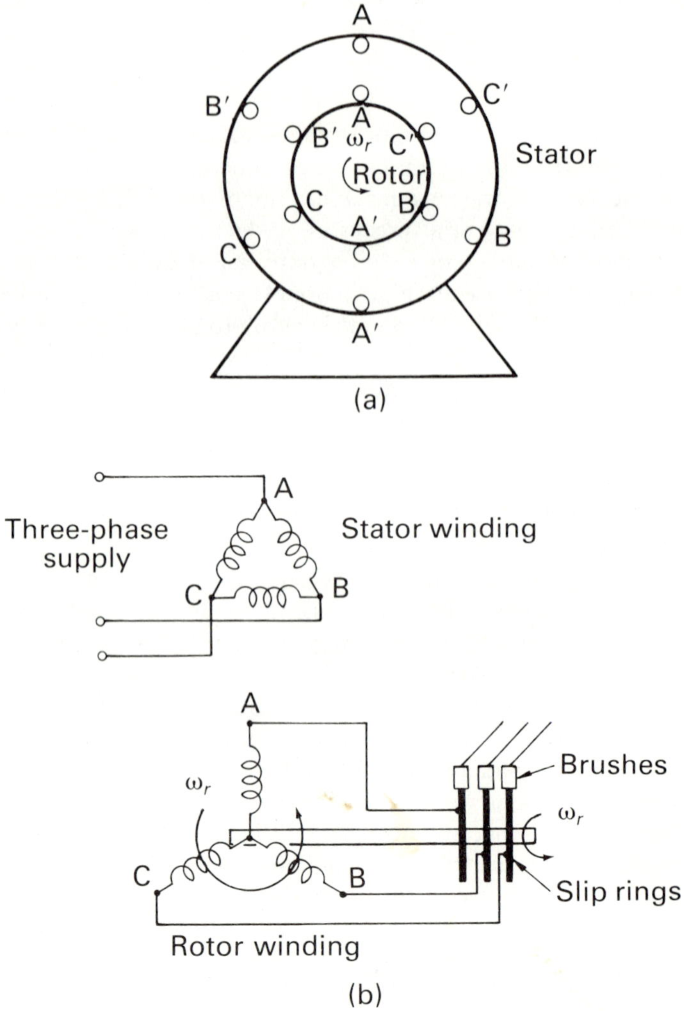

Fig 6.1 (a) Diagrammatic representation of rotor and stator coils and (b) stator and rotor winding connections.

The stator carries thee-phase currents of angular frequency ω and this will give rise to a flux density wave B_s rotating at ω in the air gap. Rotor conductors rotating at angular frequency ω_r, where $\omega_r < \omega$, will 'see' the flux wave B_s sweeping past at angular velocity $(\omega - \omega_r)$ and will therefore experience an induced e.m.f. of angular frequency $(\omega - \omega_r)$. This e.m.f. acting on the closed rotor coils will give rise to rotor coil currents of angular frequency $(\omega - \omega_r)$ and because of the symmetrical displacement of the coil sides shown in Fig. 6.1(a) the coil currents will be a three-phase set. These three-phase rotor currents will give rise to a flux density wave B_r rotating in the air gap at angular velocity $(\omega - \omega_r)$ *with respect to the rotor coils.* Since the rotor coils already have an angular velocity of ω_r due to their rotation, B_r will have a total angular velocity *with respect to the stator* of $\omega_r + (\omega - \omega_r)$, i.e. ω. The two flux density waves B_s and B_r therefore have the same angular velocity and this is a necessary condition for an average torque to be developed between the sources of B_s and B_r.

The resultant flux density wave $(B_s + B_r)$ sweeping past the stator coils at angular frequency ω will generate an e.m.f. which, if we neglect the small stator coil impedances, must be equal to the applied terminal voltage $\overline{V}_p$. This is precisely the same situation as we found in the stator coils of the synchronous machine in Section 5.3. Since the normal mode of operation is for the motor to be run with constant applied terminal voltage then the resultant flux wave $(B_s + B_r)$ must be constant. The space vectors representing $\mathbf{B}_s$, $\mathbf{B}_r$ and $(\mathbf{B}_s + \mathbf{B}_r)$ are shown in Fig. 6.2(a) while in Fig. 6.2(b) the corresponding flux density distribution waves are shown. Let us call the resultant flux density wave B_0 so that we may draw analogies between B_0 and the resultant core flux density ϕ_0 in a transformer. Let us further split B_s into two components as follows.

$$\mathbf{B}_s = \mathbf{B}_0 + \mathbf{B}$$

But

$$\mathbf{B}_s + \mathbf{B}_r = \mathbf{B}_0$$

$$\therefore \quad \mathbf{B}_r = -\mathbf{B} \tag{6.1}$$

Equation (6.1) shows us that the rotor currents set up a flux density wave B_r which cancels part of the stator flux density wave B_s leaving the resultant flux density wave B_0 in the air gap. There is an exact parallel here to the secondary currents in a transformer cancelling part of the magnetising effect of the primary currents to leave a resultant core flux ϕ_0. Whereas the transformer coil e.m.f's are set up by the time variation of a stationary flux distribution $N\,[\mathrm{d}\phi_0/\mathrm{d}t]$ the induction motor coil e.m.f's are set up by the movement of a time invariant flux density distribution $(B_0 lv)$. (It should be remembered that $B_0 lv$ is equivalent to $N\,[\mathrm{d}\phi_0/\mathrm{d}t]$, when $\delta\phi_0/\delta t$ is zero.)

With reference to Fig. 6.2 the rate at which the flux density vectors, and

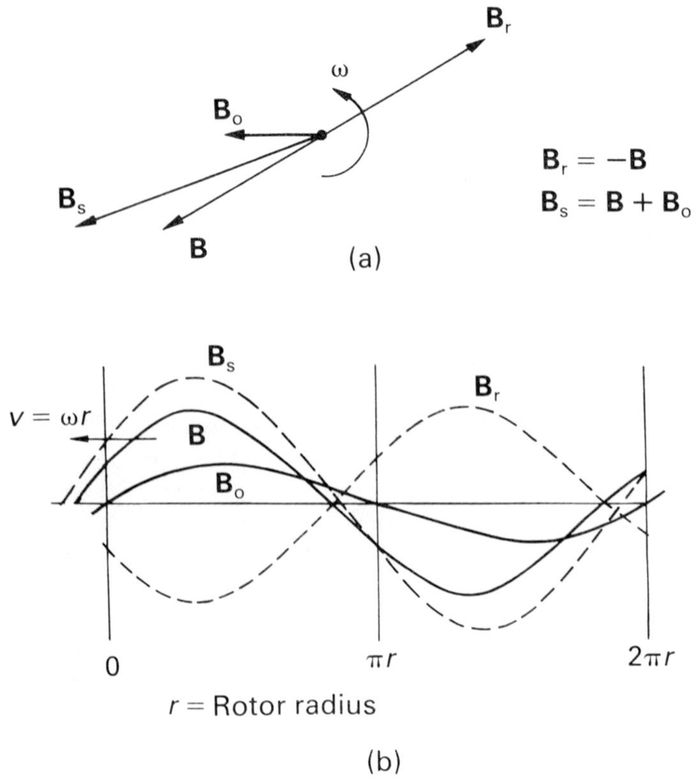

Fig. 6.2 (a) Flux-density vectors and (b) flux density distribution waves in the air-gap.

the distributions which they represent, are rotating will depend on the frame of reference. A stator coil will see the vectors rotating at ω while a rotor coil will see them rotating at $(\omega - \omega_r)$. The corresponding values for v in Fig. 6.2(b) would be ωr and $(\omega - \omega_r)r$, where r is the radius at the air gap. *The phase relationship between the vectors is the same no matter which reference frame is being used.*

The coil e.m.f's and currents can now be related to the flux density distribution vectors in the same way as was done in Section 5.2 on the synchronous machine. Fig. 6.3(a) shows coil AA′ on the rotor at the instant when the current in this coil is a maximum in the direction shown. $\mathbf{B}_r$, the rotor derived flux density distribution vector due to all three rotor coil currents, lies along the axis of coil AA' at this instant. The e.m.f. in coil AA′ is proportional to $B_0 v$ where $v = (\omega - \omega_r)r$ and will be at its maximum when the peak of the B_0 distribution coincides with AA′. Fig. 6.3(b) shows the phasor diagram for the instant under consideration.

If we use the subscript 'a' to denote coil AA′ variables then i_a is at its

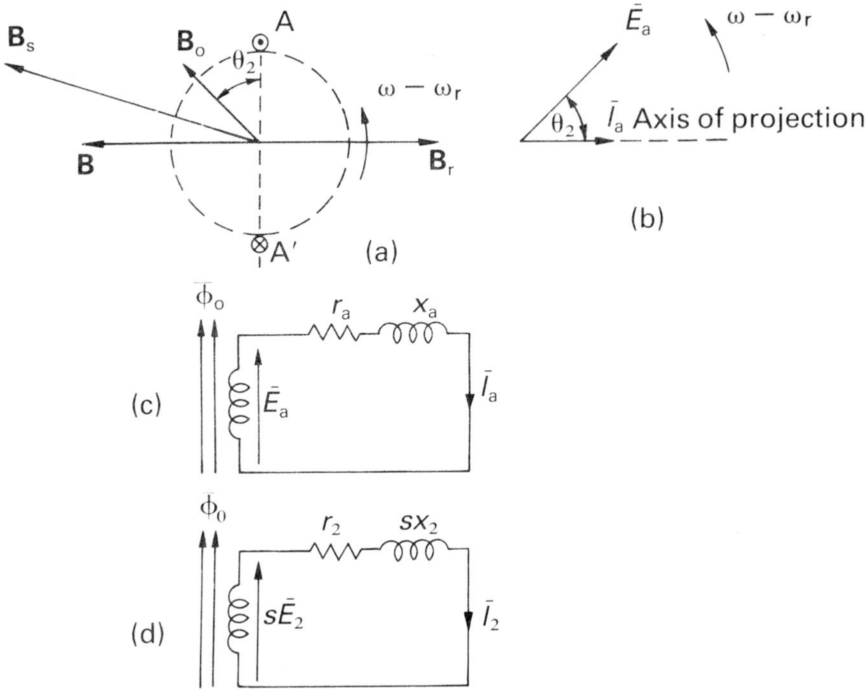

Fig. 6.3 (a) Flux density vectors as seen by rotor coil AA′, (b) phasor diagram for rotor coil AA′ (c) circuit diagram for rotor coil AA′ and (d) circuit diagram for rotor coil.

positive maximum at the instant shown with direction as shown in Fig. 6.3(a). The e.m.f. e_a was at its maximum θ_2 radians ($\theta_2/(\omega - \omega_r)$ seconds) before this instant in the same direction as that shown in Fig. 6.3(a). (Use Fleming's RH rule remembering that v is the velocity of the conductor w.r.t. the field.) The phasor diagram therefore shows $\bar{I}_a$ lying along the projection axis and $\bar{E}_a$ leading $\bar{I}_a$ by an angle θ_2. Since the rotor coils are short circuited $\bar{E}_a$ and $\bar{I}_a$ must be related by the circuit shown in Fig. 6.3(c), where r_a is the coil resistance and x_a the coil leakage reactance at the angular frequency $(\omega - \omega_r)$. The coil self-inductance and mutual inductance terms from other coils do not enter into the $\bar{E}_a, \bar{I}_a$ relationship since we are using the resultant flux density wave $\mathbf{B}_0$ as the source of $\bar{E}_a$. This is precisely the same technique as we used in the transformer equivalent circuit where the self and mutual inductance terms were eliminated as a result of using the resultant core flux $\bar{\phi}_0$. To reinforce the analogy a phasor $\bar{\phi}_0$ is shown on Fig. 6.3(c). At any instant ϕ_0 is $\int_A B_0 \, dA$ taken over the rotor surface between coil sides A and A′. The time variation in this case is produced by the relative movement between the flux wave $\mathbf{B}_0$ and the coil AA′.

The quantity $x_a = (\omega - \omega_r)l_a$ will vary as ω_r varies and it is convenient to express both x_a and $\overline{E}_a$ in terms of their respective values measured with $\omega_r = 0$ and with the resultant flux distribution vector $\mathbf{B}_0$ the same amplitude as in Fig. 6.2(a). With the rotor stationary ($\omega_r = 0$) let $\overline{E}_a = \overline{E}_2$, and $x_a = \omega l_a = x_2$. (These values are often referred to as 'standstill' values.) The angular frequency of the phasors $\overline{E}_a$, $\overline{\phi}_0$ and $\overline{I}_a$ under this condition will be ω. At any other value of ω_r,

$$x_a = (\omega - \omega_r)l_a = \frac{(\omega - \omega_r)}{\omega}(\omega l_a)$$

$$= s x_2$$

where the 'slip'

$$s = (\omega - \omega_r)/\omega.$$

$$\overline{E}_a = \frac{d\overline{\phi}_0}{dt} = j(\omega - \omega_r)\,\overline{\phi}_0.$$

When

$$\omega_r = 0,\ \overline{E}_a = \overline{E}_2.$$

$$\therefore\ \frac{\overline{E}_a}{\overline{E}_2} = \frac{\omega - \omega_r}{\omega} = s,$$

hence

$$\overline{E}_a = s\overline{E}_2 \qquad (6.2)$$

Fig. 6.3(d) shows a phase of the rotor with the e.m.f. $s\overline{E}_2$ and leakage reactance sx_2 replacing $\overline{E}_a$ and x_a. The rotor phase resistance has been labelled r_2 rather than r_a and $\overline{I}_2$ replaces $\overline{I}_a$. r_2 is not a function of slip and $\overline{I}_2 = s\overline{E}_2/(r_2 + jsx_2)$.

Fig. 6.4 shows what is happening in the stator coil AA′ at the same instant as for Fig. 6.3.

Fig. 6.4(a) repeats the flux distribution vectors shown in Fig. 6.3(a) but AA′ is now the stator coil for phase A and this coil will see the flux vectors rotating at ω and not $(\omega - \omega_r)$ as we had for the rotor coil. The stator phase A coil has been shown in the same position as the rotor phase A coil in Fig. 6.3(a). This simplifies our analysis and is indeed a situation which arises periodically but it is not always the case. For instance, after one cycle of the supply current $\overline{I}_a$ in Fig. 6.4(c), the flux vectors will again be in the positions shown in Fig. 6.3(a) and Fig. 6.4(a) but the rotor phase A coil will only have completed ω_r/ω of one revolution. On the other hand the rotor coil currents are $[(\omega - \omega_r)/\omega] \times 2\pi$ further on in their cycle so that the position of $\mathbf{B}_r$ w.r.t. the rotor coil AA′ axis has moved on by $(\omega - \omega_r)/\omega$ of one revolution. Hence the flux vector $\mathbf{B}_r$ has completed one revolution and is back in the same position as in Fig. 6.3(a). This no longer coincides with the flux axis of the rotor coil AA′.

With the $\overline{E}_a$ and $\overline{I}_a$ directions defined as in Fig. 6.4(c), i.e. *in opposite directions* in the winding, the corresponding phasor diagram is shown in

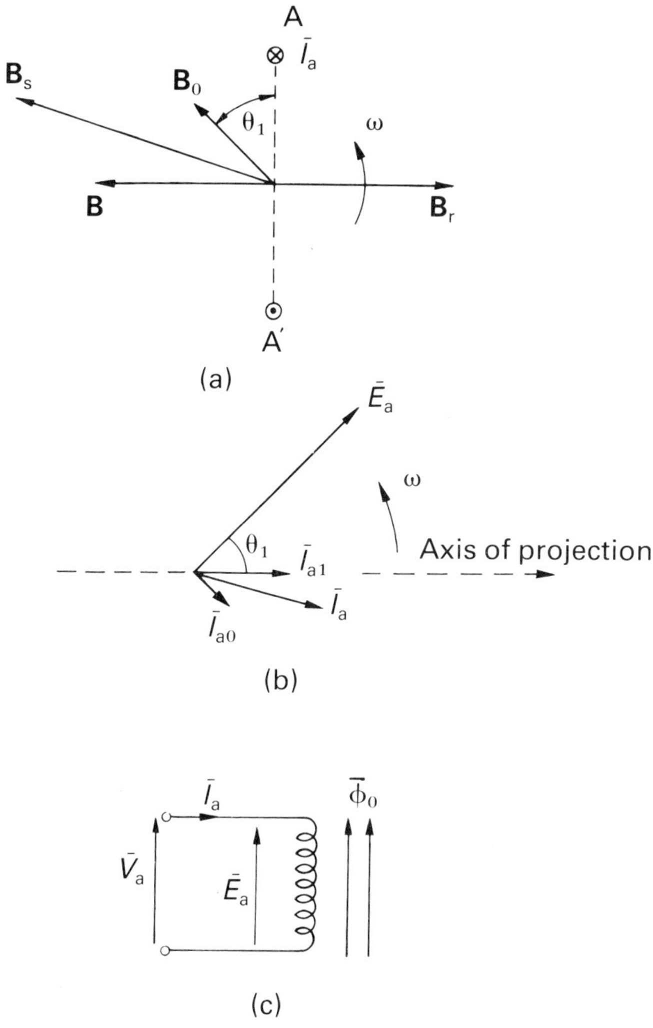

Fig. 6.4 (a) Flux density vectors as seen by stator coil AA′, (b) phasor diagram for rotor coil AA′ and (c) circuit diagram for stator coil AA′.

Fig. 6.4(b). $\bar{E}_a$ is at right angles to the B_0 position and $\bar{I}_a$ is in the opposite direction to the $\mathbf{B}_s$ direction. Let us now split the stator current $\bar{I}_a$ into two components $\bar{I}_{a0}$ and $\bar{I}_{a1}$. $\bar{I}_{a0}$, along with $\bar{I}_{b0}$ and $\bar{I}_{c0}$ in their respective phase windings, produces $\mathbf{B}_0$ and $\bar{I}_{a1}$ along with $\bar{I}_{b1}$ and $\bar{I}_{c1}$ produces **B**. $\bar{I}_{a0}$ is equivalent to the magnetising current $\bar{I}_o$ in the transformer of Chapter 3 and $\bar{I}_{a1}$ is equivalent to $\bar{I}_1$, that component of the primary current which cancels the magnetising effect of the secondary current $\bar{I}_2$. In Fig. 6.4 $\bar{I}_{a0}$ is in the opposite direction to $\mathbf{B}_0$, and $\bar{I}_{a1}$ in the opposite direction to **B**. The angle between $\bar{E}_a$ and $\bar{I}_{a1}$ is θ_1 which is the same angle as we had in Fig. 6.3(b) between the rotor coil e.m.f. and current (θ_2).Drawing further on the analogy with the transformer we develop the stator coil equivalent circuit as shown in Fig. 6.5. The phase voltage is shown as $\bar{V}_p$, the e.m.f. $\bar{E}_a$ as $\bar{E}_1$, $\bar{I}_{a1}$ as $\bar{I}_1$ and $\bar{I}_{a0}$ as $\bar{I}_0$. $\bar{I}_0$ is shown flowing in a separate hypothetical winding (see Fig. 3.3) and, along with similar currents in phases B and C, produces $\mathbf{B}_0$ and hence ϕ_0. ϕ_0 in our simple model is again $\int_A B_0.\mathrm{d}A$ taken over the stator surface between coil sides A and A′. It remains now to produce a link between the stator variables $\bar{E}_1$, $\bar{I}_1$ and the rotor variables $s\bar{E}_2$ and $\bar{I}_2$.

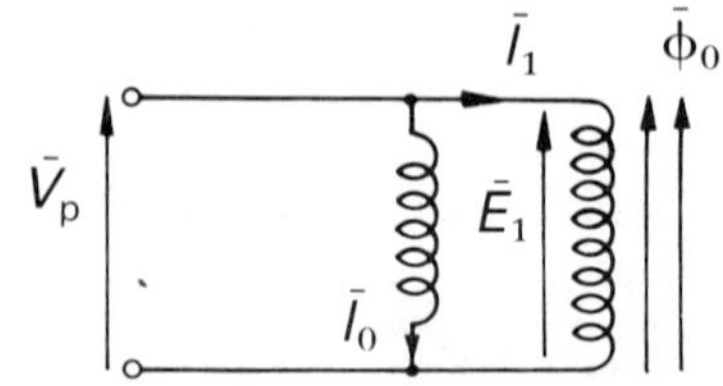

Fig. 6.5 Circuit diagram for a stator coil showing magnetising winding carrying I_0.

In the case of the transformer we were able to say $\bar{E}_1 = N_1\ (\mathrm{d}\phi_0/\mathrm{d}t)$, $\bar{E}_2 = N_2\ (\mathrm{d}\phi_0/\mathrm{d}t)$ and hence $\bar{E}_1/N_1 = \bar{E}_2/N_2$.

Can we do the same thing in this case ? Our simplified model consists of one turn coils in the centre of the phase band. If these were replaced by N_1 and N_2 concentrated turns at the positions shown for AA′ in Fig. 6.3(a) and Fig. 6.4(a) respectively then with the rotor held stationary $\bar{E}_1/\bar{E}_2 = N_1/N_2$ would be correct. On the other hand, if the rotor is held stationary with the axis of rotor coil AA′ not coincident with the axis of stator coil AA′ there would be a phase difference between $\bar{E}_1$ and $\bar{E}_2$. The scalar relationship $E_1/E_2 = N_1/N_2$ would, however, still hold. When we consider the actual phase A stator winding which will be N_1 turns distributed over the stator surface available to phase A, the expression for E_1 must be modified to take account of the phase difference between the e.m.f's in each of the N_1 turns. This was shown in Section 4.3 to modify the expression to $E_1 = k_{d1}N_1$ $(\mathrm{d}\phi_0/\mathrm{d}t)$ where k_{d1} was called the distribution factor and was dependent on the coil geometry. The expression for E_2 would be similarly modified to $E_2 = k_{d2}N_2\ (\mathrm{d}\phi_0/\mathrm{d}t)$ such that

$$E_1/E_2 = k_{d1}N_1/k_{d2}N_2 \qquad (6.3)$$

where the rotor and stator coils have the same geometry this simplifies to $E_1/E_2 = N_1/N_2$.

The currents $\bar{I}_1$ and $\bar{I}_2$ can be related by the fact that they produce rotating flux waves $\mathbf{B}$ and $\mathbf{B}_r$ respectively and these waves have equal amplitudes. It was left as an exercise for the reader in Chapter 4 to show that the amplitude of the fundamental component in the rotating flux wave set up by a distributed three-phase winding was propotional to $k_d I$. Hence the relationship between the amplitudes of I_1 and I_2 is

$$k_{d1}\, I_1 N_1 = k_{d2}\, I_2 N_2$$

$$\therefore \quad \frac{I_1}{I_2} = \frac{k_{d2} N_2}{k_{d1} N_1} \tag{6.4}$$

This will again simplify for similar rotor and stator windings to

$$\frac{I_1}{I_2} = \frac{N_2}{N_1}$$

In practical machines this simplification is never possible since there are good reasons, not covered in this text, why the coil geometry factors should be different for rotor and stator. It will always be true, however, that equations (6.3) and (6.4) can be stated in the form

$$\frac{E_1}{E_2} = \frac{K_1 N_1}{K_2 N_2}, \qquad \frac{I_1}{I_2} = \frac{K_2 N_2}{K_1 N_1} \tag{6.5}$$

where K_1 and K_2 are constants depending on stator and rotor coil geometry. Let us therefore define an 'effective' turns ratio as

$$\frac{N_1'}{N_2'} = \frac{K_1 N_1}{K_2 N_2} \tag{6.6}$$

We now have two separate circuits representating the stator phase A coil as in Fig. 6.5 and the rotor phase A coil as in Fig. 6.3(b). The common link is the flux ϕ_0 but, the difficulty is that the rate of change of ϕ_0 is different in the two circuits except at standstill. This gives rise to phasors of different frequency in the two coils and we cannot therefore superimpose the two phasor diagrams of Figs. 6.3(b) and 6.4(b). Can we find an equivalent 'stationary' coil to represent the effects of the rotor coil as 'seen' by the stator coil ? After all the rotor derived flux wave B_r is seen by the stator coil to be moving at the same velocity as the stator derived flux wave B_s. The stator coil has no way of knowing whether B_r is produced by another set of stationary coils carrying currents of angular frequency ω or by a set of rotating coils carrying currents of angular frequency $s\omega$. If such a hypothetical coil exists how do we account for the mechanical power in an equivalent circuit consisting of two sets of stationary coils?

Looking at the power division in the actual machine will help in our quest for the equivalent stationary coil. The power per phase available at the stator is

$$P = E_1 I_1 \cos\theta_1 \tag{6.7}$$

The power per phase consumed as heat in the rotor coils is

$$P_r = I_2^2 r_2 = sE_2 I_2 \cos\theta_2 \tag{6.8}$$

Now from equations (6.5) and (6.6) we can show that

$$E_2 I_2 = E_1 I_1$$

and hence

$$P_r = sE_2 I_2 \cos\theta_2 = sE_1 I_1 \cos\theta_2 \tag{6.9}$$

But from the phasor diagrams of Figs 6.3(b) and 6.4(b) we saw that $\theta_2 = \theta_1$

$$\therefore\ P_r = sE_1 I_1 \cos\theta_1 = sP \tag{6.10}$$

The remaining $(1 - s)P$ of the stator power must be consumed in doing mechanical work. Fig. 6.6 shows the power division between the stator and rotor circuits and the output shaft (per phase)

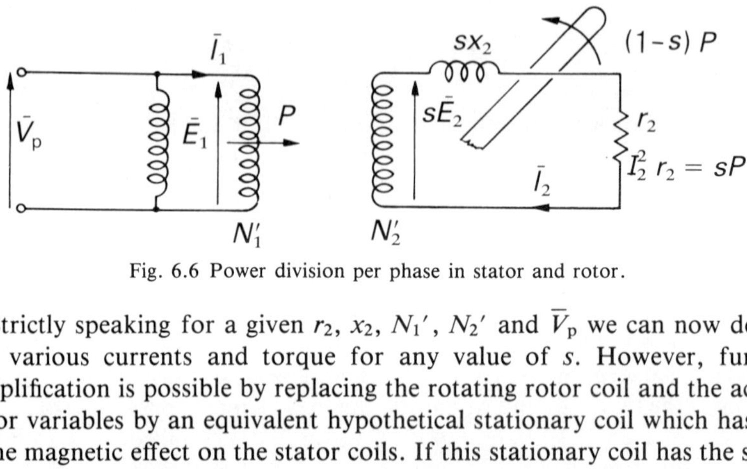

Fig. 6.6 Power division per phase in stator and rotor.

Strictly speaking for a given r_2, x_2, N_1', N_2' and $\bar{V}_p$ we can now derive the various currents and torque for any value of s. However, further simplification is possible by replacing the rotating rotor coil and the actual rotor variables by an equivalent hypothetical stationary coil which has the same magnetic effect on the stator coils. If this stationary coil has the same number of effective turns per phase (N_2') as the actual coil then the amplitude of the coil current must be the same in both hypothetical and actual coil circuits, i.e. $\hat{I}_2$. On the other hand, because the new coil is stationary the angular frequency of its current must be ω in order to produce $\mathbf{B}_r$ rotating at ω and there can be no mechanical power output. Furthermore, the coil e.m.f. will be $\bar{E}_2$ (standstill) at angular frequency ω so all of the power leaving the stator (P) must now be absorbed in the new coil electrical circuit. The actual coil absorbs only $sP = I_2^2 r_2$ and since the amplitude of I_2 remains constant the new coil can absorb P only if we change the coil resistance to r_2/s. This change is consistent with the need to keep I_2 constant while the coil e.m.f. and leakage reactance have changed to E_2 and x_2 respectively. The new coil power factor angle θ_2 has not changed since the relative positions of $\mathbf{B}_0$ and $\mathbf{B}_r$ must remain as they were for the actual rotor coil. (Remember that $\bar{E}_2$ is at right angles to the line of $\mathbf{B}_0$ and $\bar{I}_2$ is colinear with $\mathbf{B}_r$.)

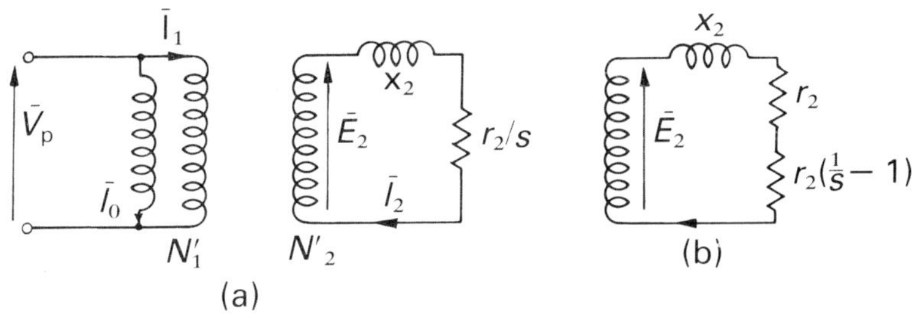

Fig. 6.7 (a) Equivalent circuit per phase and (b) modification to rotor circuit to separate heat loss and mechanical power.

Fig. 6.7(a) shows the equivalent circuit per phase with all phasors now having the same frequency. A minor modification to the new coil representing the rotor coil effects is shown in Fig. 6.7(b) where r_2/s is split into two components r_2 and $r_2\ [(1/s)-1]$. This maintains the identity of the heat loss in the rotor coil $(I_2^2 r_2 = sP)$ and the mechanical work $(I_2^2 r_2\ ((1/s)-1) = (1-s)P)$. If we now incorporate the stator coil impedance and iron loss and refer 'secondary' quantities to the 'primary' we have the simple equivalent circuit shown in Fig. 6.8. The primed quantities indicate rotor variables and

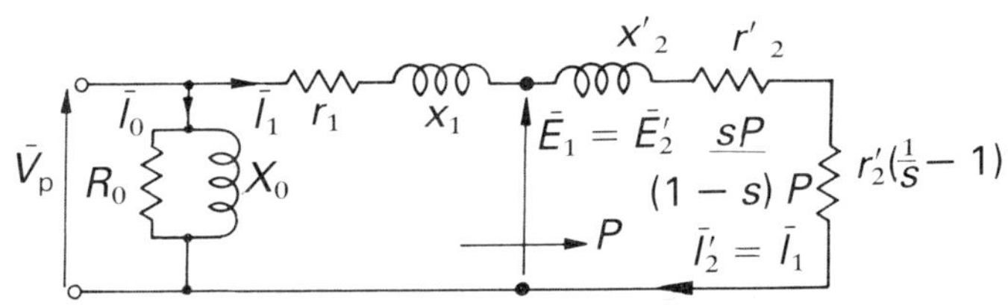

Fig. 6.8 Equivalent circuit per phase referred to stator.

parameters referred to the stator. In comparison to Fig. 6.6 the actual rotational power $\omega_r T$ in the real rotor coil has been replaced by an equivalent power in a resistor $r'_2((1/s)-1)$ in a hypothetical stationary coil.

The interested reader can check the equivalence of Fig. 6.8 and Fig. 6.6 by the following experiment on a wound rotor induction motor. Measure the amplitude and phase of the stator current over a range of applied load torques. Record also the value of slip for each load point. Now lock the rotor to prevent rotation and connect variable resistors of suitable rating into each rotor phase via the slip-rings. Vary these resistors (keeping all three equal) over a range of values and record the resistor value and the amplitude and phase of the stator coil current. Take care not to exceed the rated currents for any length of time. The value of s which corresponds to R, the added rotor resistance per phase, can be found when r_2 is known by $R = r_2 [(1/s) - 1]$. Comparison can now be made between the stator currents for a given s in the two experiments.

6.3 INDUCTION MOTOR TESTS

The equivalent of the transformer open circuit test would be for $s = 0$, i.e. the rotor going round at ω. The induction motor cannot drive itself at ω since there will always be friction and windage against which the rotor needs to develop a torque. In order to develop such a torque the rotor must rotate at slightly less than ω but provided there is no applied load torque the value of s is small and we can neglect $\bar{I}_1$ in Fig. 6.8. The no load test consists therefore of running the motor at rated voltage with no applied load torque. Input voltage, current and power are measured and the elements of the magnetising circuit (R_0, X_0) are calculated from these readings. R_0 will now include losses due to windage and friction as well as iron losses and will therefore be proportionately smaller than for a transformer of the same rating as the motor under test.

The series elements of the equivalent circuit are then found by locking the rotor and applying a reduced stator voltage sufficient to cause rated current in the stator windings. Measurement of input volts, amps and power allow calculation of $(r_1 + r_2')$ and $(x_1 + x_2')$ in Fig. 6.8. The reduced voltage, usually about 50 per cent rated, means that the iron path does not get into saturation. This will increase the value of X_0 and reduce the iron losses. R_0 will increase for this latter reason and also because at standstill there are no windage and friction losses. I_0 can therefore be neglected in comparison to I_1. s is 1 for a stationary rotor. r_1 and r_2' can be separated by a d.c. test on the stator winding to find r_1 but there is no test whereby x_1 and x_2' can be separated. Usually a reasonable estimate for a wound rotor is that $x_1/x_2' = r_1/r_2'$.

A load test is conducted by applying shaft torques within the capabilities of the motor under test and an output torque versus speed characteristic is obtained. The shape of this characteristic can then be checked against the curve computed from the equivalent circuit derived in the previous two tests.

6.4 TORQUE/SPEED CHARACTERISTIC

From the equivalent circuit of Fig. 6.8 the power available for conversion to mechanical work is $I_1^2 r_2' [(1/s) - 1]$ per phase. The total developed torque T_e is therefore given by

$$T_e = \frac{3}{\omega_m} I_1^2 \, r_2' \left(\frac{1}{s} - 1\right) \tag{6.11}$$

where ω_m is the shaft angular velocity. If the stator winding is a two-pole winding, i.e. one set of three-phase windings occupying 360° of the stator periphery then

$$\omega_m = \omega(1 - s),$$

where ω is the angular frequency of the stator phase currents. In the case of a p-pole machine then using equation (5.7), ω_m is related to ω by

$$\omega_m = \frac{2\omega(1 - s)}{p} \tag{6.12}$$

$$\therefore \; T_e = \frac{3p}{2\omega(1 - s)} I_1^2 r_2' \frac{(1 - s)}{s}$$

$$= \frac{3p}{2\omega} \frac{I_1^2 r_2'}{s} = \frac{K I_1^2 \, r_2'}{s} \tag{6.13}$$

where K is a constant for a given machine and supply frequency. Substituting for I_1 from the equivalent circuit of Fig. 6.8 gives

$$T_e = K \frac{V_1^2 r_2'/s}{\left(r_1 + \frac{r_2'}{s}\right)^2 + \left(x_1 + x_2'\right)^2} \tag{6.14}$$

When s is the only variable in equation (6.14) then T_e (max) can be found by differentiating to be:

$$T_e(\text{max}) = \pm K V_1^2 / 2(\sqrt{(r_1^2 + (x_1 + x_2')^2)} + r_1) \tag{6.15}$$

The condition giving rise to equation (6.15) is that

$$r_2'/s = \pm\sqrt{(r_1^2 + (x_1 + x_2')^2)} \tag{6.16}$$

and this is the condition for maximum power transfer into a load resistor r_2'/s from a source of impedance $r_1 + j(x_1 + x_2')$. Maximum developed torque occurs when the total power into the equivalent stationary rotor coil is a maximum. The total power is the sum of the power available for conversion $[(1 - s)P]$ and the power lost as heat in the coil resistance (sP).

Many texts on induction motors make a further approximation in the equivalent circuit which considerably simplifies the algebra but as we shall see in Example 6.1 considerably decreases the accuracy of solution for values of s greater than s at full load. This approximation consists of neglecting the stator impedance $(r_1 + jx_1)$ in the circuit of Fig. 6.8. The expression for T_e simplifies to

$$T_e = KV_1^2 \frac{r_2'}{s}\left(\frac{(r_2')^2}{s^2} + (x_2')^2\right)^{-1}$$

$$= \frac{KV_1^2 s\, r_2'}{(x_2')^2}\left(\left(\frac{r_2'}{x_2'}\right)^2 + s^2\right)^{-1} \tag{6.17}$$

Hence

$$T_e = k\,\frac{sa}{s^2 + a^2} \tag{6.18}$$

where $a = r_2/x_2 = r_2'/x_2'$.

and $T_e(\max) = \pm k/2$, when $s = \pm a$, i.e.

$$\pm s = r_2/x_2 = r_2'/x_2' \tag{6.19}$$

Example 6.1

A three-phase, 440 V, 19 kW, 50 Hz 8-pole induction motor has its rotor and stator windings star-connected. No-load and locked rotor tests gave the following results

No load	440 V	10 A	1520 W
Locked rotor	220 V	55 A	7600 W

The stator phase resistance is 0.4 Ω and the effective turns ratio stator/rotor is 2.5/1.

Estimate the parameters of the equivalent circuit and hence calculate the maximum developed torque $T_e(\max)$ and the speed at which it occurs when the motor is run from a 440 V, 50 Hz, three-phase supply.

Calculate the line current and developed torque at 710 r.p.m. and at starting. Compare these answers with those achieved when the stator impedance is neglected.

From the no-load test:

$$R_0 = (440/\surd 3)^2/(1520/3) = \underline{127.4\,\Omega}$$

$$\cos\theta_0 = 1520/(\surd 3 \times 440 \times 10) = 0.199$$

$$\therefore\ \theta_0 = 78.5^\circ,\ \sin\theta_0 = 0.98$$

$$\therefore\ x_0 = (440/\surd 3)/(10 \times 0.98) = \underline{25.92\,\Omega}$$

From the locked rotor test neglecting I_0 for reasons explained earlier, on page 152

$$I_1 = 55\text{ A and}$$

$$I_1^2(r_1 + r_2') = \frac{7600}{3} = 2533.3\text{ W}$$

$$\therefore\ r_1 + r_2' = \frac{7600}{3 \times 55^2} = \underline{0.837\,\Omega}$$

$$x_1 + x_2' = \sqrt{\left[\left(\frac{220}{\sqrt{3} \times 55}\right) - 0.837^2\right]}$$

$$= 2.152\ \Omega$$

For maximum developed torque

$$\frac{r_2'}{s} = \sqrt{[\,r_1^2 + (x_1 + x_2')^2\,]}$$

$$= \sqrt{(0.16 + 4.63)} = 2.19$$

$\therefore$ from equation (6.14)

$$T_e(\max) = \left(\frac{3 \times 8}{2 \times 2\pi \times 50}\right) \times \left(\frac{440^2}{3}\right) \times 2.19\Big/\left((0.4 + 2.19)^2 + 2.15^2\right)$$

$$= \underline{476.4\text{ Nm}}$$

The slip at which T_e (max) occurs is given by

$$s = \frac{r_2'}{2.19} = \frac{0.837 - 0.4}{2.19} = \underline{0.2}$$

The speed corresponding to this slip is

$$\omega_m = \frac{2\omega(1 - s)}{p} = \frac{2 \times 2\pi \times 50}{8} \times 0.8 = \underline{62.83\text{ rad/s}}$$

Converted to r.p.m. this gives a shaft speed of $\dfrac{62.83 \times 60}{2\pi} = \underline{600\text{ r.p.m.}}$

At 710 r.p.m., $s = \dfrac{750 - 710}{750} = \underline{0.053}$

$$\therefore\ \bar{I}_1 = (440/\sqrt{3}) \angle 0\Big/\left[\left(0.4 + \frac{0.437}{0.053}\right) + j2.15\right]$$

$$= 254 \angle 0 / (8.64 + j2.15)$$

$$= \underline{28.5 \angle 14.0 \text{ A}}$$

$$\therefore \bar{I} = \bar{I}_0 + \bar{I}_1 = 10 \angle -78.5 + 28.5 \angle -14$$

$$= \underline{34 \angle -29.4 \text{ A}}$$

From equation (6.13)

$$T_e = \frac{3p}{2\omega} I_1^2 \frac{r_2'}{s} = \left(\frac{3 \times 8}{2 \times 2\pi \times 50}\right) \times 28.5^2 \times \frac{0.437}{0.053}$$

$$= \underline{255.8 \text{ Nm}}$$

At starting

$$s = \frac{750 - 0}{750} = 1$$

$$I_1 = \frac{440}{\sqrt{3}} / \sqrt{[(0.837)^2 + 2.15^2]} = \underline{110 \text{ A}}$$

$$\therefore T_e = \frac{3p}{2\omega} I_1^2 r_2' = \underline{202 \text{ Nm}}$$

Assuming that

$$\frac{x_1}{x_2'} = \frac{r_1}{r_2'}, \quad \text{then } x_2' = 1.12 \ \Omega$$

If we neglect $r_1 + jx_1$ in the calculations then at

$$s = 0.053$$

$$\bar{I}_1 = (440/\sqrt{3}) \angle 0 \Big/ \left(\frac{0.437}{0.053} + j1.12\right)$$

$$= \underline{30.53 \angle -7.73 \text{ A}}$$

$$\therefore T_e = \frac{3p}{2\omega} \cdot I_1^2 \frac{r_2'}{s} = \underline{293 \text{ Nm}}$$

$$\text{At } s = 1, \ I_1 = \left(\frac{440}{\sqrt{3}}\right) / (0.437 + j1.12) = \underline{211.3 \text{ A}}$$

$$\therefore T_e = \frac{3p}{2\omega} I_1^2 \frac{r_2'}{s} = \underline{745 \text{ Nm}}$$

Comparison of the values for I_1 and T_e show that neglecting stator impedance produces errors of the order of 10 per cent at low values of slip ($s \leq 0.05$) but at high value of slip ($s \geq 0.1$) the errors become unacceptable. The reason for this lies in the fact that as long as the total series impedance is dominated by r_2'/s then errors are small.

* * *

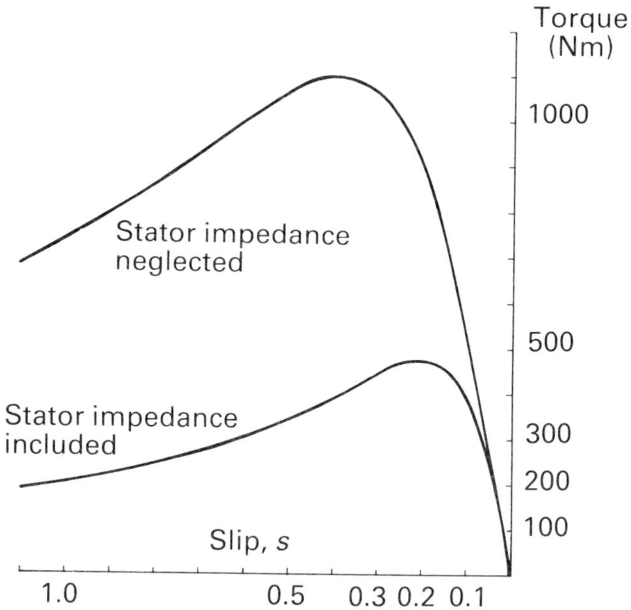

Fig. 6.9 Torque versus slip curves showing the effect of neglecting stator impedance.

Fig. 6.9 shows the effect on T_e of neglecting the stator impedance for the machine used in Example 6.1. Over the normal load range of slip values, i.e. $0 \leq s \leq 0.1$ the approximation is reasonable but should not be used elsewhere. As pointed out in the example it is the ratio r_2'/s which affects the range of validity not just s and for a different machine where r_2' was larger in comparison to $\sqrt{r_1^2 + (x_1 + x_2')}$ then the approximation would be valid over a larger slip range.

For a given machine equation (6.15) tells us that T_e(max) is independent of r_2' and equation (6.16) shows that the slip at which T_e(max) occurs will change if r_2' changes. These relationships indicate one possible way in which the speed and torque of a wound rotor induction motor may be altered to suit an application. Fig. 6.10(a) shows a set of variable resistors connected to the slip rings of a wound rotor induction motor. r_2 and hence r_2' can be adjusted as these resistors are varied and the corresponding effect on the torque/speed curves is as shown in Fig. 6.10(b).

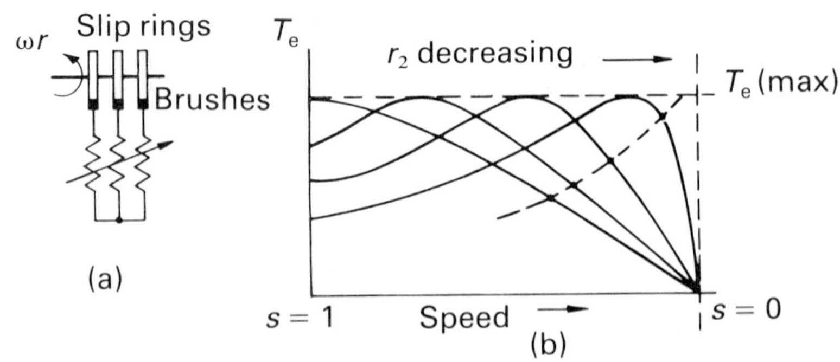

Fig. 6.10 (a) Adding resistance to rotor coils via slip-rings and (b) effect on torque/speed curves of varying r_2.

The dotted line on Fig. 6.10(b) shows the torque/speed characteristic of a possible load and the intersections give the speed at which the load will be driven as r_2 is varied. The disadvantage of this type of speed control is that unless the load torque falls off with speed the drive will be inefficient at low speeds. If the load torque remains substantially constant as the speed drops then from equation (6.13),

$$T_e = KI_1^2 \frac{r_2'}{s}$$

we can see that as s and r_2' increase such that r_2'/s is constant then I_1 is constant. The power consumed in r_2' must therefore increase as r_2' increases while the output power $\omega_m T_e$ is decreasing due to decreasing ω_m. Increased heat loss in the variable resistors connected to the rotor coils is therefore accompanied by a decrease in output shaft power with both effects contributing to a decrease in efficiency.

On the other hand increased rotor resistance is often used during starting to increase the starting torque as shown in Fig. 6.10(b) and decrease the

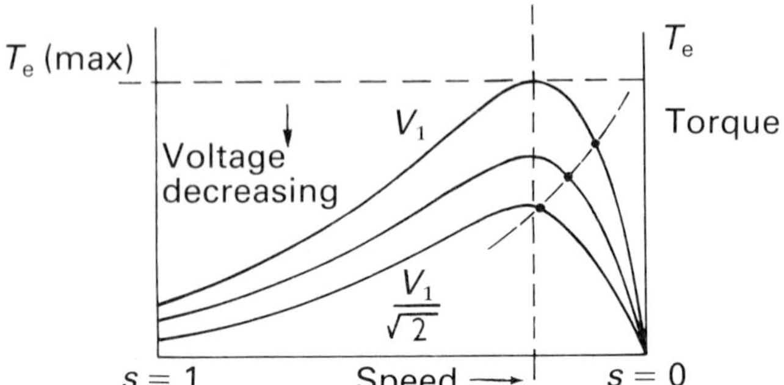

Fig. 6.11 Torque/speed curves for different values of applied voltage.

starting current. The slip rings are subsequently short-circuited when the motor is up to normal running speed.

A limited degree of speed control is available by altering V_1. T_e is proportional to V_1^2 (equation (6.14)) and dropping V_1 by $1/\sqrt{2}$ will halve the torque available at a given speed. Fig. 6.11 shows speed control for the same load as shown in Fig. 6.10(b) with constant r_2 and varying V_1. Decreasing V_1 will decrease the iron loss (V_1^2/R_0) and if the motor is running on light load this may offset any decrease in efficiency due to increasing slip. In general, however, speed decrease by either r_2 or V_1 variation will result in decreasing efficiency.

Example 6.2

The motor given in Example 6.1 is to drive a constant torque load of 300 Nm. Neglecting any changes in the mechanical losses due to changes in speed calculate the minimum added rotor resistance required for the machine to run up to speed. At what speed will the machine run if the added resistance is (a) left in and (b) subsequently shorted out ? Compare the efficiency of running at the speed calculated for condition (a) when this speed is achieved by (c) adding rotor resistance and (d) reducing the applied voltage with no added rotor resistance.

From the solution to Example 6.1 the starting torque with no added rotor resistance is 202 Nm. The motor cannot therefore start against a load of 300 Nm.

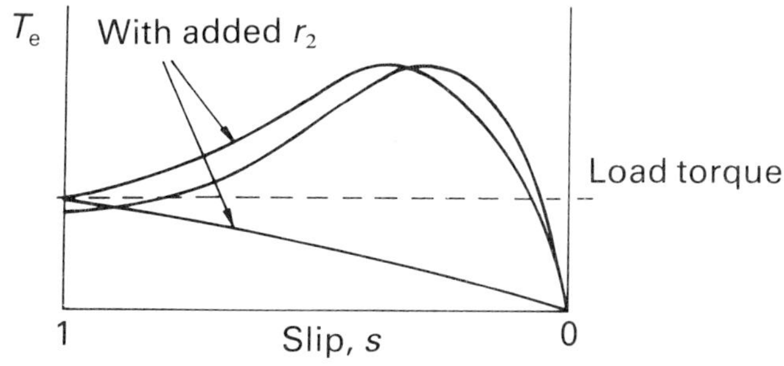

Fig. 6.12 Torque/slip curves for Example 6.2.

Using equation (6.14) and putting $s = 1$

$$300 = \frac{3p}{2\omega} \cdot \frac{V_1^2}{(r_1 + r_2')^2 + (x_1 + x_2')^2} r_2'$$

$$300 = \left(\frac{3 \times 8}{2 \times 2\pi \times 50}\right) \times \left(\frac{440}{\sqrt{3}}\right) \times \frac{r_2'}{(0.4 + r_2')^2 + (2.15)^2}$$

$$= \frac{2465\ r_2'}{0.16 + 0.8r_2' + (r_2')^2 + 4.62}$$

$$(r_2')^2 + 0.8r_2' + 4.78 - \frac{2465}{300} r_2' = 0$$

$$(r_2')^2 - 7.42\ r_2' + 4.78 = 0$$

$$r_2' = (7.42 \pm \sqrt{[7.42^2 - 4 \times 4.78]})/2$$

$$= \underline{6.7 \text{ or } 0.712\ \Omega}$$

The larger value of r_2' corresponds to the lower curve shown on Fig. 6.12 while the smaller value gives the rising curve through 300 Nm at zero speed. The motor will not start for the former but will run up to speed for the latter.

$$\text{Hence added value of } r_2 = (0.712 - 0.437) \times \frac{1}{2.5^2}$$

$$= \underline{0.044\ \Omega}$$

(Note that r_2' has to be referred back to the rotor to get the value.) To find the speed at which the motor runs we now solve,

$$300 = \frac{2465 r_2'/s}{(r_1 + r_2'/s)^2 + 4.62}$$

and since this is the same equation as above with r_2'/s replacing r_2' we shall get

(a) $\dfrac{r_2'}{s} = 6.7$ or 0.712

but $r_2' = 0.712\ \Omega$

$\therefore\ s = 1$ (as expected) or $\dfrac{0.712}{6.7} = \underline{0.106}$

$s = 0.106$ corresponds to a speed of $750\ (1 - s) \doteqdot \underline{670 \text{ r.p.m.}}$

(b) $\dfrac{r_2'}{s} = 6.7$ or 0.712 with $r_2' = 0.437\ \Omega$

$\therefore\ s = \underline{0.065}$ or $\underline{0.614}$

$s = 0.065$ corresponds to a speed of $750\ (1 - s) \doteqdot \underline{701 \text{ r.p.m.}}$

(c) The efficiency when $r_2' = 0.712$:

$$\text{Output power} = \frac{300 \times 750\ (1-s) \times 2\pi}{60}$$

$$= \underline{21064\ \text{W}}$$

$$\text{Power lost in rotor resistance} = \frac{21064\ s}{(1-s)}$$

$$= \underline{2498\ \text{W}}$$

Power lost as iron loss, windage and friction

$$= 3V_1^2/R_0$$

$$= \frac{3 \times 440^2}{3} / 127.4$$

$$= \underline{1519\ \text{W}}$$

$$\text{Efficiency} = 21064/(21064 + 2498 + 1519)$$

$$= \underline{84\ \%}$$

(d) With $r_2' = 0.437$ and $s = 0.106$ we have to find V_1^2 from

$$300 = \left(\frac{3 \times 8}{2 \times 2\pi \times 50}\right) V_1^2 \cdot (0.437/0.106) \Bigg/ \left(\left(\left|\frac{0.437}{0.106}\right|\right)^2 + 4.62\right)$$

$$\therefore\ V_1^2 = \frac{300 \times 200\pi}{24} \times \frac{21.62}{4.12} = 4.121 \times 10^4$$

$\therefore$ Power lost in iron losses, windage and friction

$$= 3 \times V_1^2/127.4$$

$$= \underline{970.3\ \text{W}}$$

The output power and power lost in rotor resistance is as in (c) (same torque and slip)

$$\therefore\ \text{efficiency} = 21064/(21064 + 2498 + 970.3)$$

$$= \underline{85.9\ \%}$$

With reference to (d) it is true that the iron losses have dropped but the other rotational losses have not since the speed is constant. Our estimate of efficiency is therefore slightly optimistic but it will be greater than for (c).

6.5 SQUIRREL-CAGE MOTORS

Fig. 6.13 (a) A 900 kW wound rotor or slip-ring induction motor (6 poles, 985 r.p.m.) (b) Cage rotor of a 4-pole, 11 MW, 1484 r.p.m. induction motor. (Courtesy G.E.C. Large Machines Ltd.)

So far we have considered the rotor to be wound in a similar fashion to the stator. Such motors are referred to as wound rotor or slip-ring induction motors. Fig. 6.13 (a) shows a typical rotor structure.

Where there is no need to gain access to the rotor windings then a much simpler, more robust rotor structure is possible. In its simplest form each rotor slot carries a single conducting bar which is connected to all other rotor bars by conducting end rings. Fig. 6.13 (b) shows a typical rotor structure while Fig. 6.14 shows a development of the rotor surface indicating the position of the B_0 wave, the conductor e.m.f's and currents.

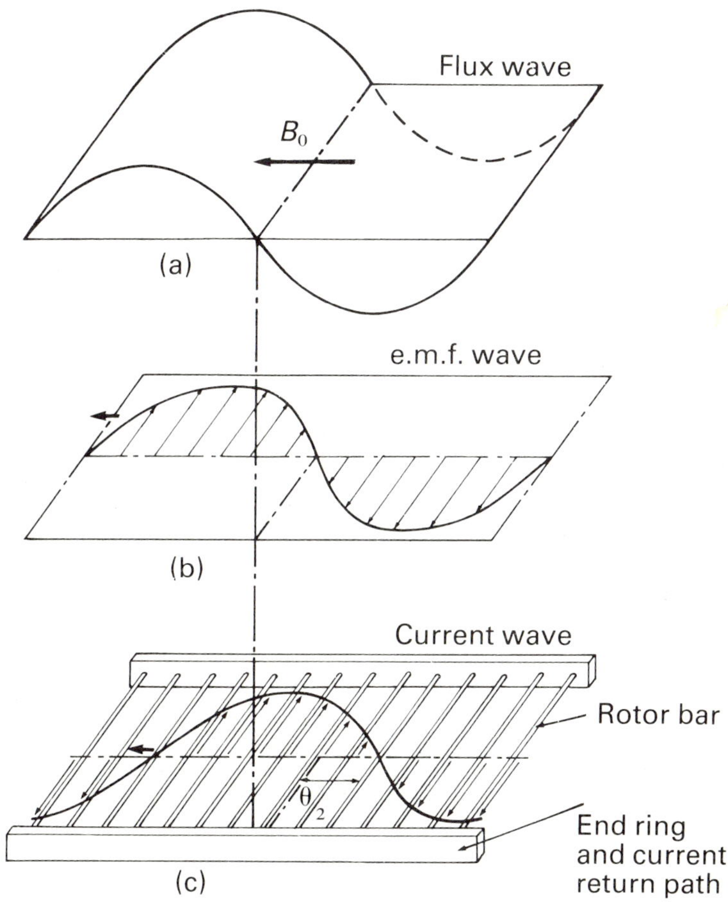

Fig. 6.14 (a) The resultant flux density wave, B_0, (b) bar e.m.f. distribution and (c) bar currents.

Note that the currents are not in phase with the e.m.f's due to the leakage inductance of the current loops. The current distribution is much closer to a sine wave than for the three-phase machine where currents in the coil sides of a given phase are equal. Here the currents can vary from one bar to the next and as m, the number of bars, becomes larger the current distribution tends towards an exact sine wave. The number of 'poles' in the machine, i.e. the number of maxima and minima of the B_0 distribution in 360 degrees mechanical, depends only on the number of sets of three-phase windings on the stator. Although the distinct phase groups on the rotor have disappeared the effects of rotor currents as seen from the stator can still be represented in terms of an equivalent rotor three-phase winding which would produce the same fundamental component in the B_r flux density distribution. The equivalent circuit referred to the stator is still a valid representation of this type of machine but we no longer have the ability to alter r_2 or x_2 by making external connections to the rotor circuit. The torque speed curve of a simple squirrel-cage motor would be similar to that shown in Fig. 6.9 and it follows that it can only be used in applications not requiring high starting torques. Small speed variations will still be possible by altering the applied voltage.

In order to improve the starting characteristics of squirrel-cage motors, designers have made use of the finite time it takes the stator flux wave to penetrate into the rotor iron. At starting and low speeds the flux remains close to the rotor surface due to the relatively high rate of change of flux in the rotor. As the rotor speed approaches that of the air gap flux-waves the flux return paths penetrate deeper into the rotor iron. Two separate cage windings, one near the rotor surface and one embedded deeper in the rotor iron, can then be used to give a combined torque/speed curve as shown in Fig. 6.15 (a). Fig. 6.15 (b) shows how the original double cage winding has evolved into a deep-slot bar which produces the same effect. The combined characteristic is based on a surface winding with a high resistance and low leakage reactance which predominates at low speed giving way to a low resistance high leakage reactance winding at high speed. The leakage reactance will increase as the rotor current path gets further away from the stator

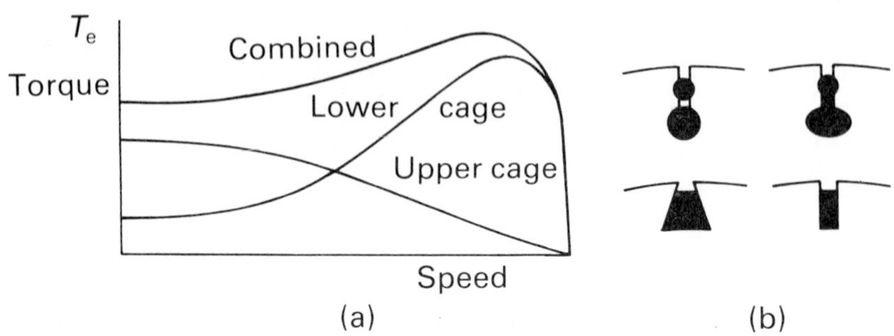

Fig. 6.15 (a) Torque/speed curve for double cage rotor and (b) methods of achieving double cage effect.

conductors. The exact shape of the torque/speed curve can be varied by altering the slot and conductor shape but in most applications a characteristic which is near rectangular in shape is suitable. Because r_2 and x_2 are now frequency dependent the equivalent circuit and test procedures are more complex.

6.6 STARTING THREE-PHASE INDUCTION MOTORS.

Wound rotor induction motors can be connected directly-on-line (DOL) with appropriate values of resistance connected into the rotor coils. These resistors serve to increase the starting torque and decrease the starting current. The resistors are subsequently decreased to zero for normal running speeds. Most squirrel cage motors are capable of DOL starting without sustaining any damage to themselves but they may draw very large surges of current from the mains giving rise to unacceptable voltage dips in the

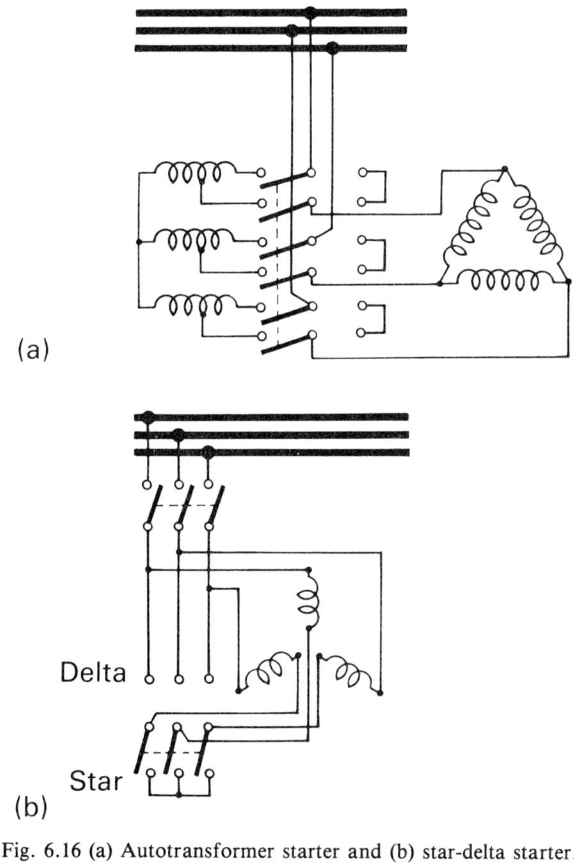

Fig. 6.16 (a) Autotransformer starter and (b) star-delta starter.

local supply network. Where the local network is 'strong', i.e. it has a very low source impedance, induction motors of several thousand horsepower (kW/0.746) may be started DOL. In general, however, above say 5 h.p. precautions may have to be taken to reduce starting current surges. Wherever possible the motor should be decoupled from the load since decreasing the load torque and inertia will decrease the run-up time. The voltage applied to the motor terminals is then reduced by either (a) star connecting a normally delta connected motor or, (b) supplying the motor via a three-phase autotransformer.

By connecting the winding in star the phase voltage is reduced by $\sqrt{3}$ in comparison to the delta connection. This will reduce the phase currents by $\sqrt{3}$ and the line currents by 3 in comparison to the delta connection. When the motor has reached approximately 80 per cent of normal running speed a special contactor reconnects the windings in delta for normal operation. This reconnection must be carried out quickly in order to avoid current surges comparable with those for DOL starting. Star-delta starting is not generally used when the supply voltage exceeds 3kV. Large motors therefore require an autotransformer to reduce starting currents. The autotransformer may have one or more fixed taps which are used progressively in the run-up procedure. Fig. 6.16 shows the stator connection for (a) autotransformer starting and (b) star-delta starting.

'Soft start' power electronic packages are also now available. The user need only push the 'Start' button and the control electronics keeps the supply current within a preset limit.

6.7 SINGLE-PHASE INDUCTION MOTORS

6.7.1 Introduction

Very few, if any, textbooks cover the single-phase motor before three-phase motors for the simple reason that three-phase motors are simpler devices around which to develop a rotating flux wave theory and a simple equivalent circuit. Indeed, prior to Ferraris' idea of representing a pulsating quantity by two contra-rotating vectors the idea of a contra-rotating field theory was not there to tempt the engineer to indiscretion. Should the reader wish to learn more about the indiscretion in question, then he must read the relevant paper by Williams and Laithwaite [3]. With the benefit of hindsight we will avoid the pitfalls and produce a correct analysis.

6.7.2 The equivalent circuit

The vast majority of single-phase motors are in reality unbalanced two-phase motors. Once up to speed some may become single phase motors when the starting winding is switched out but many continue to operate as two-phase motors. Let us first consider the operation of a balanced two-phase motor. The stator carries two windings with their magnetic axis at 90° as shown in Fig. 6.17 (a). B_a and B_b are cosinusoidally distributed around the air gap with their maximum occurring at the vector position shown. The

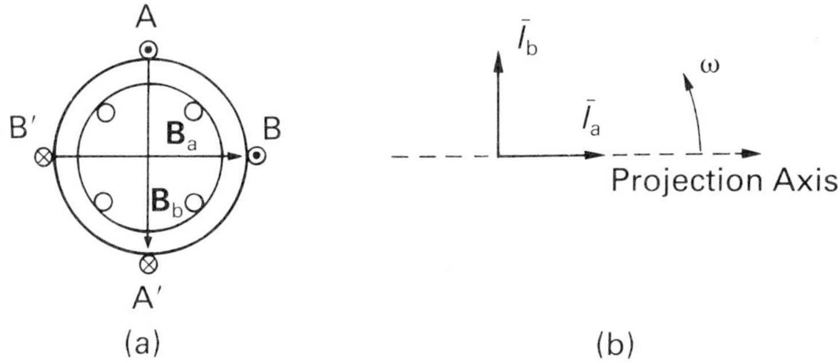

Fig. 6.17 (a) Diagrammatic representation of stator and rotor coils and flux-density vectors and (b) phasor diagram for stator coil currents.

distributed coils which AA′ and BB′ represent carry currents $\bar{I}_a$ and $\bar{I}_b$ respectively which are out of phase by 90° as shown in Fig. 6.17 (b). At the instant shown the length of B_a is at its maximum while that of B_b is zero. The resultant flux density in the air gap at a position $\theta°$ in an anticlockwise direction from the axis of B_a is

$$B_s = B_a \cos\theta + B_b \cos(90 + \theta) \tag{6.20}$$

The time variation of B_a and B_b correspond to the time variation of I_a and I_b respectively so that

$$B_s = B_a \cos\omega t \cos\theta + B_b \cos(\omega t + 90) \cos(\theta + 90)$$

and if $I_a = I_b$ this becomes

$$\begin{aligned} B_s &= B_a\,(\cos\omega t \cos\theta + \cos(\omega t + 90)\cos(\theta + 90)) \\ &= B_a\,(\cos\omega t \cos\theta + \sin\omega t \sin\theta) \\ &= B_a \cos(\omega t - \theta) \end{aligned} \tag{6.21}$$

The resultant flux density distribution B_s given by equation (6.21) can therefore be represented by a vector of length B_a rotating anticlockwise at angular frequency ω. The resultant flux density at any instant is cosinusoidally distributed about the position of B_s at that instant. We can now employ the same arguments as we used in the development of the three-phase motor equivalent circuit to arrive at the same equivalent circuit per phase for the two-phase machine. The two-phase machine requires a supply system consisting of two equal voltages V_a and V_b which are out of phase by 90°. Such voltages applied to identical windings (AA′ and BB′) will then give rise to $\bar{I}_a$ and $\bar{I}_b$ as shown in Fig. 6.17(b). The rotor winding in Fig. 6.17(a) is shown as a two-phase winding but as in the three-phase case the rotor may be of the squirrel cage type without invalidating the equivalent circuit.

Most two-phase machines are not fed from balanced two-phase voltage sources as in the foregoing analysis. Nor are the two windings AA′ and BB′ identical. This results in the maximum amplitude of the $\bar{B}_a$ vector being different to the $\bar{B}_b$ vector. The currents $\bar{I}_a$ and $\bar{I}_b$ do not form a symmetrical two-phase set, i.e. $\bar{I}_b \neq \pm j\bar{I}_a$. If $\bar{I}_b = \alpha \angle (90 - \phi)\ \bar{I}_a$ then equation (6.20) becomes

$$B_s = B_a \cos \omega t \cos \phi \ - B_b \sin(\omega t - \phi) \cos(90 + \theta)$$

$$= B_a \cos \omega t \cos \phi \ + B_b \sin(\omega t - \phi) \sin \phi$$

The amplitude relationship between B_a and B_b will be affected by the factor α in the current relationship and any inequality in the effective turns for coil AA′ and coil BB′. Both effects can be incorporated in the statement $B_b = \beta B_a$.

$$\therefore\ B_s = B_a\ (\cos \omega t \cos \theta + \beta \sin(\omega t - \phi) \sin \theta) \qquad (6.22)$$

Let us modify the right-hand side of equation (6.22) so that we get a rotating field as before plus a remainder term, i.e.

$$B_s = B_a(\cos \omega t \cos \theta + \sin \omega t \sin \theta - \sin \omega t \sin \theta$$

$$+ \beta \sin(\omega t - \phi) \sin \theta)$$

$$= B_a(\cos(\omega t - \theta) + \beta \sin(\omega t - \phi) \sin \theta - \sin \omega t \sin \theta)$$

$$= B_a(\cos(\omega t - \theta) + (\gamma \sin(\omega t - \epsilon)(\sin \theta)) \qquad (6.23)$$

where

$$\gamma = \surd\, [(\beta \cos \phi \ - 1)^2 + (\beta \sin \phi)^2]\,, \epsilon = \tan^{-1} \frac{\beta \sin \phi}{\beta \cos \phi \ - 1}$$

The first term on the right-hand side of equation (6.23) is a rotating field component as before. The second term as seen by an observer on axis θ is a pulsating component on an axis at $(90 - \theta)$. This pulsating component can, by Ferraris's principle, be split into two contra-rotating components such that

$$B_s = B_a \left(\cos \underset{\circlearrowleft}{(\omega t - \theta)} + \frac{\gamma}{2} (\cos \underset{\circlearrowleft}{(\omega t - \epsilon - \theta)} - \cos \underset{\circlearrowright}{(\omega t - \epsilon + \theta)}) \right) \qquad (6.24)$$

The various components in equation (6.24) have been marked to indicate the direction of rotation of that particular field component. Remember that θ is an angle arising from position while ϵ is an angle arising out of the phase difference between $\bar{I}_a$ and $\bar{I}_b$. If we take the anticlockwise direction as 'forward' then there are two forward components and one backward component. In order to see their spatial relationship, let us consider the instant $t = 0$. The first term has its maximum at $\theta = 0$ and a vector of length B_a lies on this axis as shown in Fig. 6.18(a).

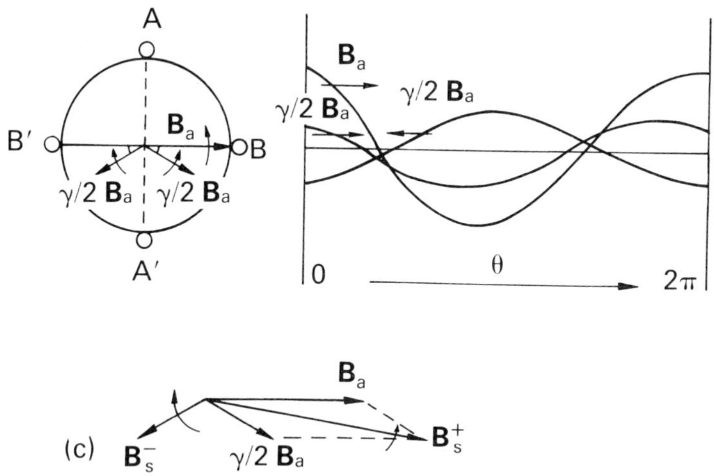

Fig. 6.18 (a) Position of flux density vectors at instant $t = 0$, (b) position of flux-density waves at instant $t = 0$ and (c) combination of flux density vectors into $\mathbf{B}_s^+$ and $\mathbf{B}_s^-$.

The second term has its maximum at $\theta = -\epsilon$ and the third term at $\theta = \epsilon + \pi$ (allowing for the negative sign). Fig. 6.18 (a) shows the position of the contra-rotating vectors at the instant chosen while Fig. 6.18 (b) shows the air-gap flux density distributions which they represent. The two forward components can be combined vectorially as in Fig. 6.18 (c) to give a single forward vector B_s^+. The single backward rotating vector is renamed B_s^-.

The unbalanced two-phase machine can now be solved in terms of two balanced two-phase machines. One of these machines has stator phase currents which give rise to B_s^+ and the other has stator phase currents which give rise to B_s^-. The net current in any stator phase will then be the sum of the phase currents in the two balanced machines and likewise the net phase e.m.f. will be the sum of the two e.m.f's. While the two machines are balanced in the sense that each machine is characterised by a single rotating flux distribution vector, we must remember that the effective number of turns on the phase windings of each machine must be the same as on the actual machine. If the ratio of the turns on the phase 'a' winding to that on the phase 'b' winding is n then in each of the two balanced machines

$$nI_a = I_b, \qquad E_a = nE_b \tag{6.25}$$

(Note that these are amplitude relationships only. There will always be a 90° phase relationship between corresponding e.m.f. and current phasors in the two phases of the balanced machines.)

For a machine running at slip s in the B_s^+ direction then the equivalent circuits for phases 'a' and 'b' of the balanced machine representing the forward direction are as shown in Fig. 6.19 (a) and (b) respectively.

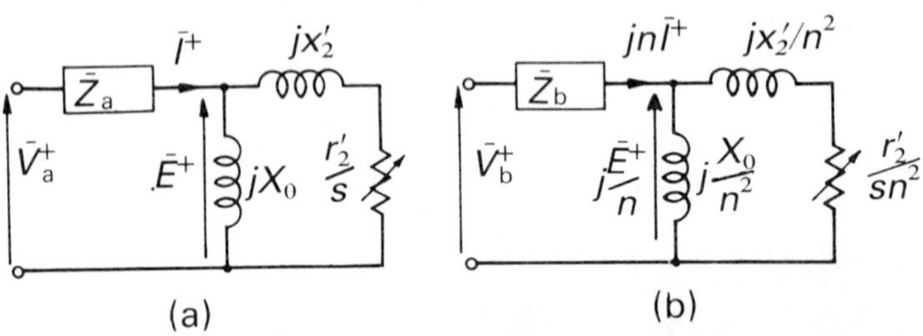

Fig. 6.19 Equivalent circuit for forward components for (a) phase a, and (b) phase b.

The significance of n can be seen to be identical to the turns ratio in a transformer in terms of common flux effects reflected into windings with differing numbers of turns. The stator impedances are seen to be different while the effects of the balanced rotor windings (probably squirrel cage) differ only in respect of the turns ratio squared.

(Note that: $\bar{V}_b^+ \neq j\bar{V}_a^+/n$).)

For the balanced machine representing the reverse direction the slip for a rotor rotating at $\omega_m(1 - s)$ in the opposite direction is $(\omega_m + \omega_m(1 - s))/\omega_m$, i.e. $(2 - s)$. The phase 'a' and 'b' equivalent circuits for the reverse direction machine are shown in Fig. 6.20 (a) and (b) respectively.

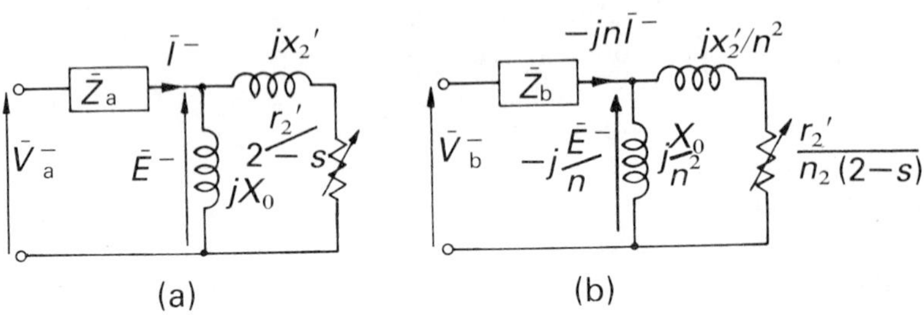

Fig. 6.20 Equivalent circuit for backward components for (a) phase a, and (b) phase b.

Note the use of $-j$ in place of j in Fig. 6.20 (b) in comparison to Fig. 6.19(b). Phase 'b' quantities now lag phase 'a' quantities as opposed to leading in the forward field case.

The air-gap flux density in the actual machine results from the simultaneous existence of the two contra-rotating vectors B_0^+ and B_0^-. (B_0 is the resultant vector arising from the interaction of B_s and B_r.) The e.m.f's and currents shown in Fig. 6.19 and 6.20 will also exist simultaneously in their respective windings such that

$$\left.\begin{aligned} \bar{I}_a &= \bar{I}^+ + \bar{I}^- & \bar{E}_a &= \bar{E}^+ + \bar{E}^- \\ \bar{I}_b &= jn\bar{I}^+ - jn\bar{I}^-, & \bar{E}_b &= j\frac{\bar{E}^+}{n} - j\frac{\bar{E}^-}{n} \end{aligned}\right\} \tag{6.26}$$

But

$$\bar{V}_a = \bar{E}_a + \bar{I}_a\bar{Z}_a,$$

and

$$\bar{V}_b = \bar{E}_b + \bar{I}_b\bar{Z}_b, \tag{6.27}$$

where $\bar{V}_a$ and $\bar{V}_b$ are the applied voltages of phases 'a' and 'b' respectively. Substitution from equations (6.26) into (6.27) and subsequent manipulation leads to equations

$$\begin{aligned} \bar{V}^+ &= \bar{E}^+ + \bar{I}^+(\bar{Z}_a + n^2\bar{Z}_b)/2 + \bar{I}^-(\bar{Z}_a - n^2\bar{Z}_b)/2 \\ \bar{V}^- &= \bar{E}^- + \bar{I}^-(\bar{Z}_a + n^2\bar{Z}_b)/2 + \bar{I}^+(\bar{Z}_a - n^2\bar{Z}_b)/2 \end{aligned} \tag{6.28}$$

These equations are derived in Appendix C where it is also shown that

$$\bar{V}^+ = \frac{1}{2}(\bar{V}_a - jn\bar{V}_b)$$

$$\bar{V}^- = \frac{1}{2}(\bar{V}_a + jn\bar{V}_b)$$

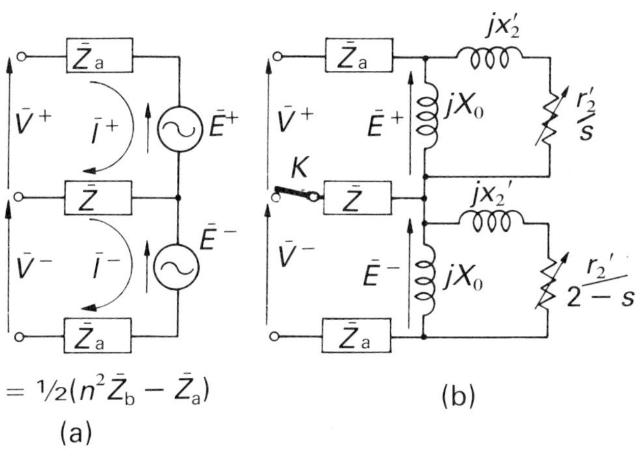

Fig. 6.21 (a) Circuit diagram for equation (6.28) and (b) equivalent circuit for unbalanced two-phase motor.

Equations (6.28) describe a circuit as shown in Fig. 6.21 (a) and in Fig. 6.21 (b) the $\bar{E}^+$ and $\bar{E}^-$ sources are replaced by the equivalent circuit elements from Fig. 6.19 (a) and Fig. 6.20 (a).

A switch K is also shown in series with $\bar{Z}$ in Fig. 6.21. While both windings 'a' and 'b' carry currents this switch is closed. Many 'single'-phase motors are started with both windings excited and winding 'b' is subsequently open circuited by a centrifugal switch when the motor has reached a significant speed. When $\bar{I}_b$ goes to zero then from equation (6.26) we can see that $(\bar{I}^+ - \bar{I}^-)$ goes to zero. This is equivalent to opening switch K in Fig. 6.21(b). With K open $\bar{I}^+$ is equal to $\bar{I}^-$ such that

$$\left.\begin{aligned} \bar{I}_a &= \bar{I}^+ + \bar{I}^- = 2\bar{I}^+ \\ \text{and} \qquad \bar{V}^+ &= \bar{V}^- = \frac{\bar{V}_a}{2} \end{aligned}\right\} \tag{6.29}$$

The equivalent circuit for a single-phase motor in terms of $\bar{V}_a$ and $\bar{I}_a$ can be derived from Fig. 6.21 (b) with K open and all impedances divided by 2. This is shown in Fig. 6.22 where a resistor R_o has also been inserted to cater for iron, windage and friction losses.
(These losses might have been accounted for by putting a resistor in parallel with each $X_o/2$.)

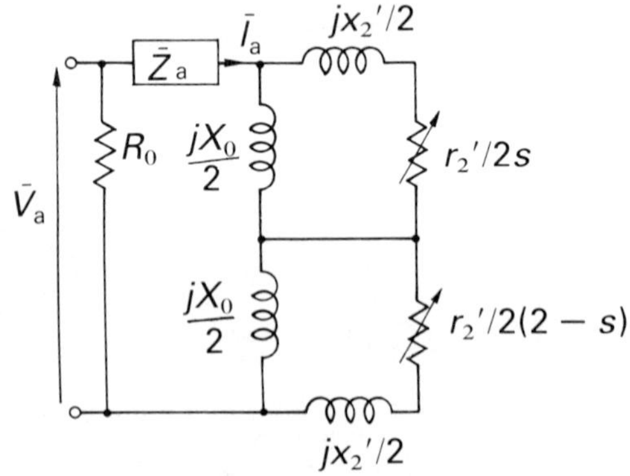

Fig. 6.22 Equivalent circuit for a single-phase induction motor.

In calculating the torque for a given value of s the reverse torque must be subtracted from the forward torque to give the net torque. With this in mind it is immediately obvious that with $s = 1$ the single phase motor of Fig. 6.22 has no starting torque. This is why such motors are fitted with two windings; a main winding which remains in circuit all the time (our phase 'a') and an auxiliary winding (our phase 'b') which is used only during starting and subsequently disconnected. The auxiliary winding can be switched out once s lies between 1 and 0 since the net torque is in the forward direction over this range.

Note that for the single phase machine the net stator flux density distribution B_s is a pulsating, not rotating field. Phase 'b' currents still flow in the rotor, however, giving rise to B_r^+ and B_r^- distributions. The B_r^+ component predominates for $0 \leq s \leq 1$.

6.7.3 Types of single-phase induction motors

The various types differ in respect of the type of auxiliary winding. Fig. 6.23 shows the two mains variations. In Fig. 6.23 (a) the phase difference between currents $\bar{I}_a$ and $\bar{I}_b$ is produced by the auxiliary winding having a much higher resistance than the main winding.

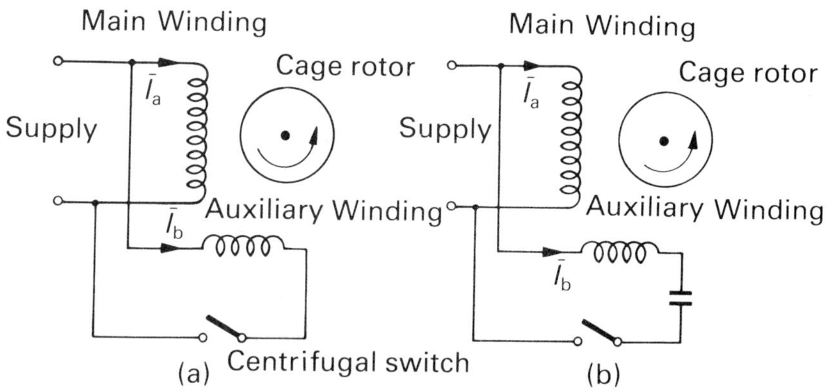

Fig. 6.23 Starting techniques for single-phase induction motors. (a) High resistance auxiliary winding and (b) series capacitor in auxiliary winding branch.

In Fig. 6.23 (b) the phase difference is largely due to the inclusion of a capacitor C in series with the auxiliary winding. In some cases the capacitor C consists of two capacitors in parallel only one of which is switched out during normal running conditions. The benefit of this arrangement is that a short duty high capacity reversible electrolytic capacitor can be used for giving a high starting torque while a much smaller capacity continuously rated non-electrolytic capacitor can be used to improve the torque available at normal running speeds. The high resistance auxiliary winding type of Fig. 6.23 (a) is usually referred to as a 'split-phase' motor while the other type is referred to as 'capacitor start' or 'capacitor start/capacitor run' motor. The capacitor motors are easily recognisable by the cylindrical electrolytic type capacitors attached to the stator frame.

Fig. 6.24 gives a rough torque/speed characteristic classification showing that the torque is greater with the auxiliary winding in circuit and that the use of capacitors increases the torque available at starting and, for the 'capacitor run' type, over the normal working slip range.

Single-phase motors of the above type will rarely exceed 5 h.p. (3.7 kW) but in special circumstances may go as high as 30 h.p. They suffer in comparison to three-phase motors in that the instantaneous torque is not constant and this may give rise to excessive noise and vibration in larger sizes.

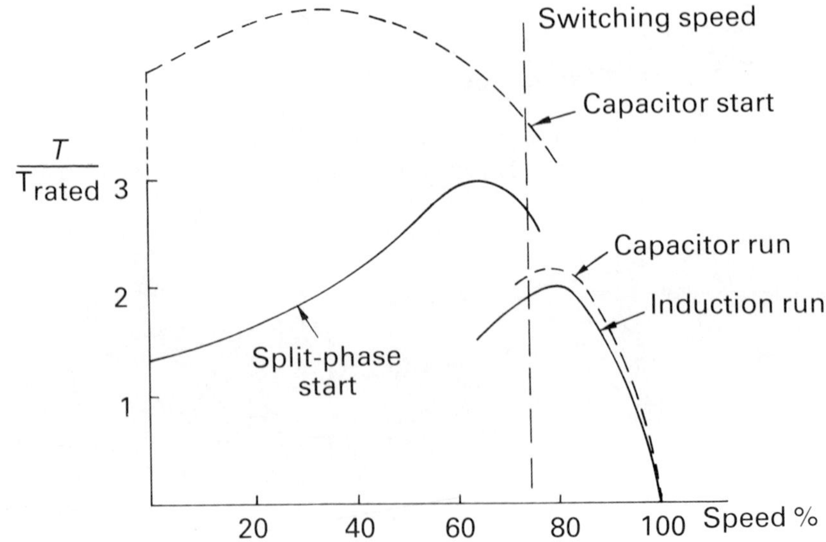

Fig. 6.24 Torque/slip curves for various types of single-phase induction motor.

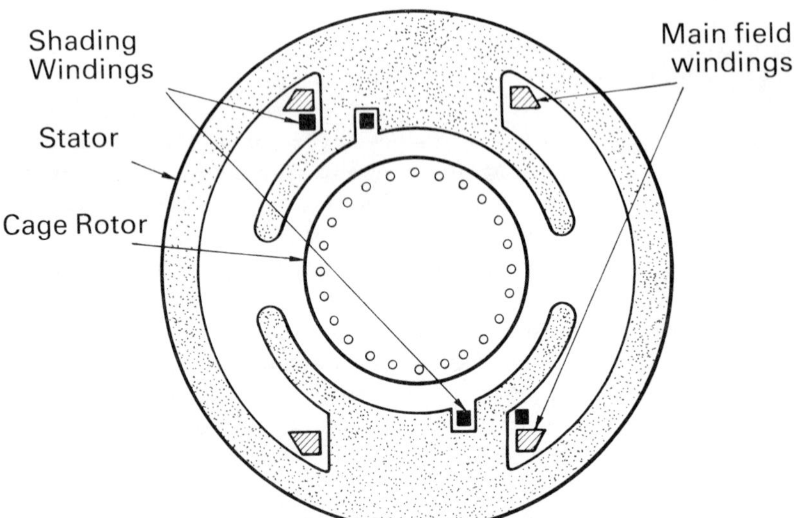

Fig. 6.25 Shaded pole motor.

For motors around 1/10 h.p. and less the auxiliary winding takes the form of a shorted turn of heavy duty copper embedded at one side of the stator pole face as shown in Fig. 6.25.

The axis of this very simple auxiliary winding is no longer 90° away from the axis of the main winding, but the analysis can still be carried out in terms of forward and backward rotating fields as before. This type of single-phase induction motor is called a 'shaded-pole' motor.

There are other non-induction type single-phase motors, notably the 'universal' motor, which will be dealt with in Chapters 7 and 8.

Example 6.3

A four-pole capacitor start/capacitor run motor is rated at 1 h.p. (0.746 kW), 240 V, 50 Hz. The values for the equivalent circuit parameters are

$$\bar{Z}_a = (4.1 + j5.6)\Omega,\ X_o = 160\ \Omega,\ r_2' = 8.24\ \Omega,\ x_2' = 4.20\ \Omega$$

$$\bar{Z}_b = (10 + j9.2)\ \Omega,\ n = 1/1.18$$

On no load the iron, windage and friction losses are 40 W.

Calculate the impedance of the run capacitor such that there is no backward field at a slip of 0.04. Find the torque at this slip a) with the run capacitor in circuit and b) with the auxiliary winding open circuited. Calculate the efficiency in both cases.

To find the value of the run capacitor we can either use Fig. 6.19 with $\bar{Z}_c$ in series with $\bar{Z}_b$, $\bar{V}_a^+ = \bar{V}_b^+ = 240 \angle 0$ and find $\bar{Z}_c$ to get the current in Fig. 6.19(b) equal to jn times the current in Fig. 6.19 (a) or we can use the circuit of Fig. 6.21 and find $\bar{Z}_c$ (included in $\bar{Z}$) to give $\bar{I}^-$ equal to zero. We shall employ the latter method leaving the former as an exercise for the reader. From Fig. 6.21 if $\bar{I}^-$ is zero then $\bar{I}^+$ can be found from Fig. 6.26 to be

$$\bar{I}^+ = \bar{V}^+/(\bar{Z}_a + \bar{Z} + \bar{Z}_f) \qquad (6.30)$$

where $\bar{Z}_f$ is the impedance of jX_0 in parallel with $r_2'/s + jx_2'$.

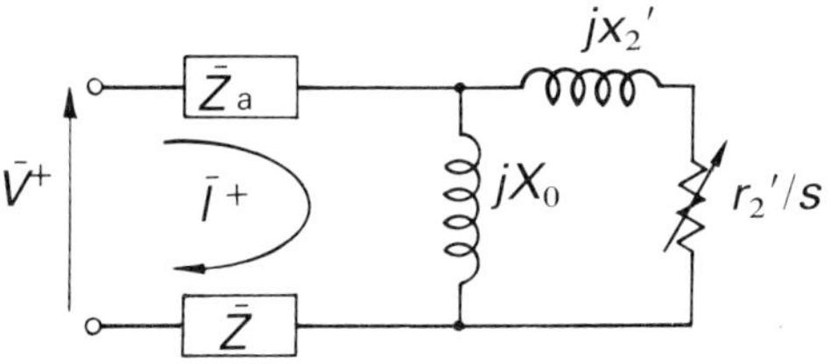

Fig. 6.26 Circuit diagram for Example 6.3.

With reference to Fig. 6.21 if $\bar{I}^-$ is to be zero then

$$\bar{V}^- = -\bar{I}^+ \bar{Z}$$

$$\therefore \quad \bar{V}^- = \frac{-\bar{V}^+ \bar{Z}}{\bar{Z}_a + \bar{Z} + \bar{Z}_f}$$

$$\therefore \quad \bar{Z}\left(1 + \frac{\bar{V}^+}{\bar{V}^-}\right) = -(\bar{Z}_a + \bar{Z}_f)$$

$$\therefore \bar{Z} = -(\bar{Z}_a + \bar{Z}_f)/(1 + \bar{V}^+/\bar{V}^-) \tag{6.31}$$

Now $$\bar{V}^+/\bar{V}^- = \frac{\tfrac{1}{2}(\bar{V}_a - jn\,\bar{V}_b)}{\tfrac{1}{2}(\bar{V}_a + jn\,\bar{V}_b)}$$

and with $$\bar{V}_a = \bar{V}_b = 240 \angle 0$$

$$\bar{V}^+/\bar{V}^- = \frac{1 - j/1.18}{1 + j/1.18} = 1 \angle -80.6$$

$$\therefore \quad 1 + \bar{V}^+/\bar{V}^- = 1.53 \angle -40.3 \tag{6.32}$$

When $s = 0.04$, $r_2'/s = 8.24/0.04 = 205\ \Omega$. x_2' is negligible in comparison and hence

$$\bar{Z}_f = \frac{j160 \times 206}{206 \times j160} = 126.4 \angle 52.7\,\Omega$$
$$= (77.5 + j99.8)\,\Omega$$

$$\therefore \quad \bar{Z}_f + \bar{Z}_a = 81.6 + j105.4 = 133.3 \angle 52\,\Omega$$

$\therefore$ from equation (6.31)

$$\bar{Z} = \frac{-133.3 \angle 52}{1.53 \angle -40.3} = -87.4 \angle 92.3\,\Omega$$
$$= \underline{(3.5 - j87.3)\,\Omega}$$

From Fig. 6.21, $\bar{Z} = \frac{1}{2}(n^2\bar{Z}_b - \bar{Z}_a)$ and in this case $\bar{Z}_b = 10 + j9.2 + \bar{Z}_c$ since $\bar{Z}_c$ is in series with the auxiliary winding (phase b).

$$\therefore \quad 3.5 - j87.3 = \frac{1}{2}\left(\frac{1}{1.18^2}(10 + j9.2) - 4.1 - j5.6 + \frac{\bar{Z}_c}{1.18^2}\right)$$

$$\therefore \quad \bar{Z}_c = 1.18^2(3.92 - j175.6)$$
$$= \underline{(5.5 - j244.5)\,\Omega}$$

and

$$C = \frac{1}{2\pi 50 \times 244.5} = \underline{13\,\mu\mathrm{F}}$$

The run capacitor will almost certainly be of the impregnated paper type and will already have a small loss resistor associated with it. The series resistance to be added will therefore be less than 5.5 Ω. Because there is no backward field at this slip the machine behaves like a balanced two-phase motor. The torque will not pulsate and the machine will be quieter.

$$\text{Torque per phase} = (I_2')^2 \frac{r_2'}{s} \frac{p}{2\omega}$$

and since there are two phases

$$\text{total torque, } T_e = \frac{2p}{2\omega} \frac{(I_2')^2(r_2')}{s}$$

$$\left(\bar{I}_2' \text{ is the current in } \frac{r_2'}{s} \text{ in Fig. 6.26.}\right)$$

Because jx_2' is negligible in comparison to r_2'/s at this slip

$$(I_2')^2 \frac{r_2'}{s} = V_f^2/r_2'/s$$

where $\bar{V}_f$ is the voltage across $\bar{Z}_f$

i.e.

$$\bar{V}_f = \frac{\bar{V}^+ \bar{Z}_f}{\bar{Z}_a + \bar{Z}_f + \bar{Z}} = \frac{120(1 - j/1.18) \times 126.4 \angle 52.7}{85.1 + j18.1}$$

$$= \frac{157.3 \angle -40.3 \times 126.4 \angle 52.7}{87 \angle 12}$$

$$= 228.5 \angle 0.4 \text{ V}$$

$$\therefore \quad T_e = \frac{2 \times p \times 228.5^2}{2\omega \times 206}$$

$$= \frac{4 \times 228.5^2}{2\pi 50 \times 206} = \underline{3.23 \text{ Nm}}$$

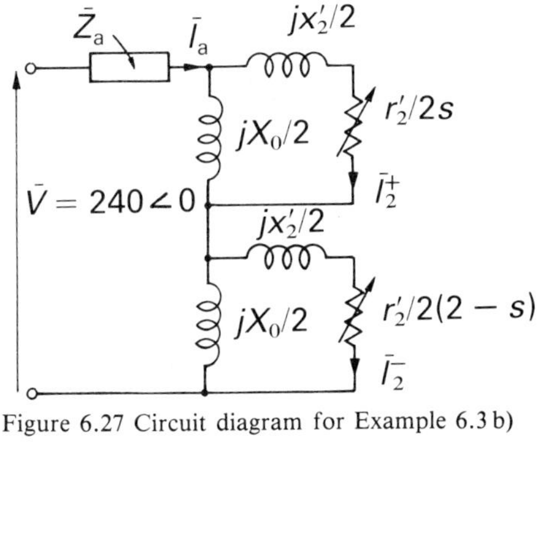

Figure 6.27 Circuit diagram for Example 6.3 b)

With the auxiliary winding out of circuit the circuit becomes as in Fig. 6.27. $\overline{Z}_f$ for the upper network in Fig. 6.27 will be half the value in the previous calculation, i.e.

$$\overline{Z}_f^+ = \frac{126.4}{2} \angle 52.7 = (38.3 + j50.3)\ \Omega$$

$$\overline{Z}_f^- = \frac{j80\overline{Z}_2^-}{\overline{Z}_2 + j80}$$

where

$$\overline{Z}_2^- = \frac{r_2'}{2(2-s)} + \frac{jx_2'}{2}$$

$$\therefore \quad \overline{Z}_2^- = \frac{8.24}{2(2-0.04)} + \frac{j4.2}{2} = (2.1 + j2.1)\ \Omega$$
$$= \underline{2.97 \angle 45\ \Omega}$$

$$\therefore \quad \overline{Z}_f^- = \frac{j80 \times 2.97 \angle 45}{2.1 + j2.1 + j80} = 2.89 \angle 46.5\ \Omega$$
$$= \underline{(2.0 + j2.09)\ \Omega}$$

$$\therefore \quad \overline{I}_a = 240 \angle 0 / [(4.1 + j5.6) + (38.3 + j50.3) + (2.0 + j2.09)]$$
$$= 240 \angle 0 / (44.4 + j57.99) = \underline{3.3 \angle -52.6\ \text{A}}$$

The power output is

$$P = (I_2^+)^2 \frac{r_2'}{2s} - (I_2^-)^2 \frac{r_2'}{2(2-s)} = \text{where } \overline{I}_2^+ \text{ and}$$

$\overline{I}_2^-$ are as shown in Fig. 6.27.

$$\overline{I}_2^- = \frac{\overline{I}_a \times j80}{j80 + 2.1 + j2.1} = j3.3 \angle -52.6 \times 0.97 \angle -88.5$$
$$= \underline{3.2 \angle -51.1\ \text{A}}$$

$$\overline{I}_2^+ = \frac{\overline{I}_a j80}{j80 + 103} = j3.3 \angle -52.6 \times 0.61 \angle -37.8$$
$$= 2.02 \angle -0.4\ \text{A}$$

$$\therefore \quad P = 2.02^2 \times 103 - 3.2^2 \times 2.1 = \underline{398.7\ \text{W}}$$

$$\therefore \quad T_e = \frac{P}{2\omega} \times p = \frac{4}{2 \times 2\pi \times 50} \times 398.7 = \underline{2.54\ \text{Nm}}$$

(Note that there is no need to consider the torque per phase in an equivalent circuit. The power levels have already been taken care of by reducing all impedances by half.)

Making the assumption that the iron, windage and friction losses are the same in both cases then the efficiency is calculated as follows.

(1) With the auxiliary winding in circuit (including run capacitor)

$$\eta = \frac{\omega_m T_e}{\omega_m T_e + 40 + I_a^2 R_a + I_b^2 R_b + \text{rotor copper loss}}$$

$$\bar{I}_a = \bar{I}^+ + \bar{I}^- = \bar{I}^+$$

$$= \bar{V}^+/(\bar{Z}_a + \bar{Z}_f + \bar{Z}) = \frac{157.3 \angle -40.3}{87 \angle 12}$$

$$= \underline{1.8 \angle -52.3\,\text{A}}$$

$$\bar{I}_b = jn\bar{I}^+ - jn\bar{I}^- \qquad = (j/1.18)1.8 \angle -52.3$$

$$= \underline{1.53 \angle 37.7\,\text{A}}$$

$$\therefore \quad I_a^2 R_a = 1.8^2 \times 4.1 = 13.3\,\text{W}$$

$$I_a^2 R_b = 1.53^2(10 + 5.5) = 36.4\,\text{W}$$

(R_b includes the resistance of run capacitor)

Rotor copper loss $= 2 \times (I_2')^2 r_2 = 20.3$ W.

$$\omega_m T_e = \frac{2\omega(1-s)}{p} . \times 3.23 = 487\,\text{W}$$

$$\therefore \qquad \eta = 487/(487 + 13.3 + 36.4 + 20.3 + 40) = \frac{487}{597} = \underline{81.6\%}$$

(Check: input power $= Re[\bar{V}_a(\bar{I}_a + \bar{I}_b)] + 40 = 594$ W.)

(2) With the auxiliary winding out of circuit

$$\bar{I}_a = 3.3 \angle -52.6\,\text{A}$$

$$\omega_m T_e = \frac{2\omega(1-s)}{p} \times 2.54 = \underline{383\,\text{W}}$$

Input power $= 240 \times 3.3 \times \cos 52.6 + 40 = 521$ W.

$$\therefore \quad \eta = \frac{383}{521} = \underline{73.5\%}$$

These efficiencies will drop as the slip increases.

6.7.4 Applications

Whenever the torque/speed characteristic of the induction motor is suitable then it will be the cheapest drive available. If the application can tolerate a speed variation of up to 5 per cent from no load to full load and the starting torque is of the same order of magnitude as the full load torque, then the induction motor is a suitable drive. The squirrel cage machine will be used where there are no special starting torque requirements or a need to abstract power from the rotor in order to run at low speeds (see Chapter 9). Otherwise a wound rotor machine is required.

Single-phase motors are widely used in domestic applications up to about 1 kW, e.g. fridge motors, washing machine pumps, fans, etc.

At the moment the largest three-phase induction motors are of the order of 25 MW, four or six-pole and if a larger power is required then a synchronous motor must be employed.

6.7.5 Conclusions

Starting from the rotating field set up by three-phase currents in a three-phase winding, we have derived an equivalent circuit for a wound rotor (slip-ring) induction motor. This circuit may also be used for squirrel cage motors at least in so far as the equivalent circuit referred to the stator is concerned.

The equivalent circuit is very similar to that derived for a transformer in Chapter 3 except that there is a secondary resistance the value of which changes with slip.

Speed control by added rotor resistance and reduced voltage are investigated and worked examples are given.

Finally the single-phase induction motor equivalent circuit is derived and used in a worked example.

PROBLEMS

1 A three-phase 240 V, 6 h.p. (6×0.746 kW) 50 Hz, six-pole, star connected slip-ring induction motor gave the following test results:

No load	240 V	5.25 A	400 W
Locked rotor	120 V	25 A	2000 W

Calculate the line current, power factor and efficiency at a speed of 932 r.p.m.

Calculate the maximum torque, maximum power output and the slip values at which these occur.

Ans. 16.1 A, 0.917, 80%; 768 Nm, 6.51 kW, 0.22, 0.169.

2 A 30 h.p. (30×0.746 kW) 400 V three-phase, 50 Hz, six-pole delta connected squirrel cage motor has the following equivalent circuit parameters:

Stator phase resistance 0.5 Ω, rotor phase resistance referred to stator 0.39 Ω, total leakage reactance per phase referred to stator 4.92 Ω. At no

load with 400 V applied the line current is 10 A at 0.2 p.f. lagging. Estimate the starting current and torque for DOL starting.
Ans. 149 A, 71.8 Nm.

3 For the motor in Example 6.3, calculate the torque, supply current and power factor, and efficiency at a slip value of 0.1.
Ans. 6.45 Nm, 5.09 A, 0.97 lagging, 76.5% (run capacitor in circuit).

7

D.C. Machines

7.1 INTRODUCTION

In 1831 Faraday first demonstrated an elementary d.c. generator using a rotating conducting disc in a undirectional field. This has since become known as the Faraday disc experiment. The magnetic field in this experiment was parallel to the axis of rotation. Machines with a magnetic field in the radial direction soon followed and by 1833 the basic prototype of today's radial field d.c. machines including Ampère's 'commutator' had been demonstrated. These early machines were soon found to be reversible, i.e. they could be used either as generators or motors. Local d.c. supply networks emerged and d.c. machines had undergone significant technological development before the introduction of a.c. supply networks and machines in the early twentieth century. Thereafter a.c. machines gradually took over from d.c. machines in most applications. The development of 'power electronics', first of all around the thyristor and now also around the transistor, has further eroded the domain of the d.c. machine in the past twenty years. Despite all this the d.c. machine, particularly the d.c. motor, is still alive and well. The automobile industry has assured the small d.c. motor of continued existence and applications requiring precise speed control over a wide range in either direction, possibly at constant torque, are still better served by a d.c. motor than by an a.c. motor plus a variable frequency power supply.

In this chapter we shall briefly consider the armature winding layout and the commutation process and then from there develop the basic equations of performance from which the output characteristics of d.c. machines can be computed. We shall draw heavily on the material in Chapters 2 and 4 and the student is advised to read these chapters prior to proceeding with this chapter.

7.2 HOMOPOLAR AND HETEROPOLAR MACHINES

Fig. 7.1(a) shows a Faraday disc-type machine. We need not concern ourselves with the coil structure which produces the field flux B_s. A d.c.

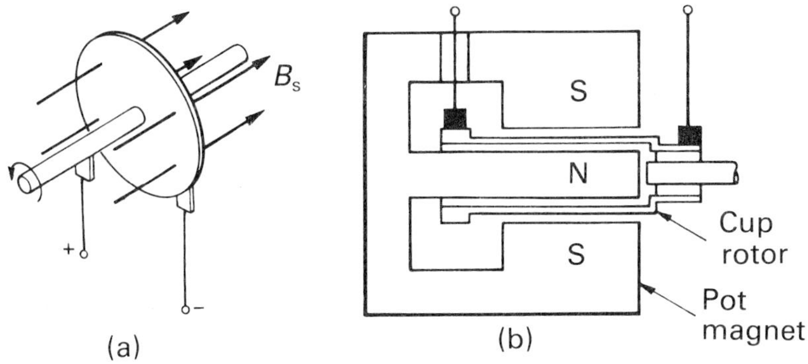

Fig. 7.1 (a) Axial flux Faraday disc machine and (b) radial flux homopolar machine.

voltage E is produced between a pair of brushes rubbing on the shaft and the disc rim. A more practical form of homopolar machine is shown in Fig. 7.1(b) where a cup rotor runs inside a pot magnet with brushes bearing on opposite ends of the cup surface. In both machines the flux direction seen by an element of the rotor surface does not change with rotation, hence the name homopolar.

When a current flows between the brushes the disc or cup experiences a circumferentially directed force giving rise to a torque or counter torque. It is interesting to note that there is no net force between the source of B_s and the current carrying circuit. The current carrying circuit crosses the flux in two regions. Once in the disc or cup and again in the fixed return lead. The forces between the field and current in these two regions are equal and opposite in their effect on the field source. The force interaction in the homopolar machine is actually between the two parts of the current carrying circuit, i.e. the fixed leads and the element in motion and not between the field source and the armature currents. Interested readers will find a more detailed explanation in the classic text by E. G. Cullwick [4].

Homopolar machines, even at high speed, generate low voltages due to their short active path length between brushes and are only useful in very low voltage high current applications.

The disc type machine of Fig. 7.1(a) would be called an axial flux machine because the field flux is in the direction of the axis of rotation. The cup-type machine of Fig. 7.1(b) would be called a radial flux machine with the field flux at right angles to the direction of rotation.

Axial and radial flux heteropolar machines are shown in Fig. 7.2(a) and (b) respectively. In such machines an element on the rotor surface will experience changes in flux direction as it moves from one pole to the next. The conducting paths on the surface of these rotors must be separated and interconnected into an armature 'winding' in order to pick up an e.m.f. If the rotors were made of a continuous conducting material as for the homopolar

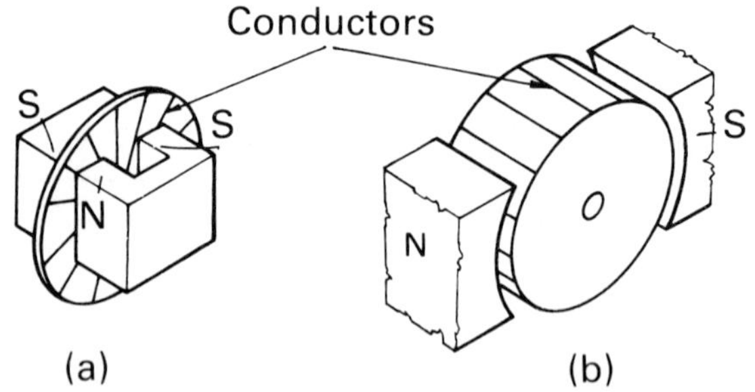

Fig. 7.2 Heteropolar machines. (a) Axial flux, and (b) radial flux.

machines then current loops (eddy currents) would flow in the rotor surfaces resisting motion and generating heat. Electrical output is achieved by windings of the type given in Chapter 4 and in order to get a d.c. output voltage the windings work in conjunction with a commutator as shown in Fig. 4.12. The windings given in Chapter 4 apply to radial flux machines only although it is not difficult to work out their axial machine equivalent. We shall confine our arguments to the more common radial flux machines.

7.3 E.M.F. AND TORQUE EQUATIONS

Fig. 7.3 shows the simple 12 conductor armature winding of Fig. 4.12(b) with the brushes replaced by point contacts which are just in contact with one edge of a commutator segment. The point contact although impractical will remove for the moment computational complications arising from the finite time occupied by the commutation process with brushes of finite width. The effects of the latter will be considered in a later section dealing specifically with the commutation process.†

The instantaneous e.m.f. between the brushes is the sum of the instantaneous e.m.f's in the conductors connected in series between the brushes. In Fig. 4.12(b) we saw that the current splits into two paths at each brush. Likewise there are two parallel paths through this armature winding which will lead from one brush to the other. Provided the flux density distribution is symmetrical and the slot spacing is even the e.m.f's in the two paths will be equal. Any inequality would give rise to large circulating currents in the winding. The paths for the instant shown are either: (1) conductors 6, 11, 4, 9, 2, 7, or (2) conductors 1, 8, 3, 10, 5, 12.

If e_n represents the e.m.f. in the nth conductor and the positive direction for e_n is from bottom to top in Fig. 7.3(a) then the instantaneous e.m.f.

† The conductors have also been rearranged into two-conductor groups with one conductor above the other. Each two-conductor group would lie in a slot in the rotor surface.

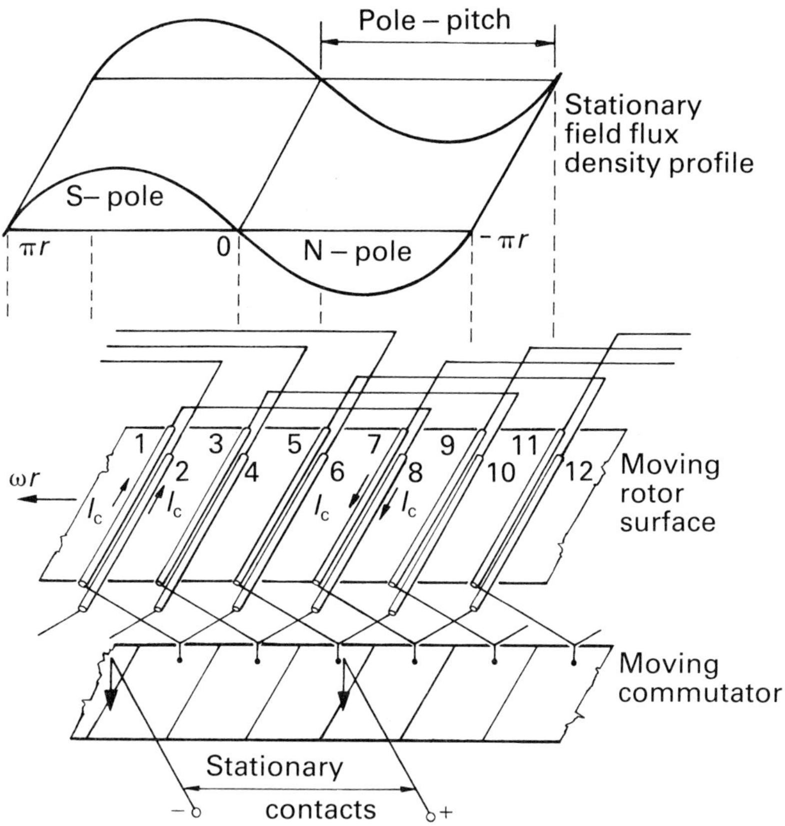

Fig. 7.3 12 conductor winding with 2 conductors per slot and brushes replaced by sliding point contacts.

between the brushes is

$$\sum e = e_1 - e_8 + e_3 - e_{10} + e_5 - e_{12}$$

Now because of the symmetrical flux density distribution and even slot spacing we have

$$e_8 = -e_2, \qquad e_{10} = -e_4, \qquad e_{12} = -e_6$$

Hence, $\sum e = e_1 + e_2 + e_3 + e_4 + e_5 + e_6$ or by similar arguments for the other path

$$\sum e = -(e_7 + e_8 + e_9 + e_{10} + e_{11} + e_{12})$$

The instantaneous e.m.f. $\sum e$ can therefore be found by summing the e.m.f's in the conductors under one pole. The e.m.f. in any conductor e_n is

$$e_n = B_n l \omega r$$

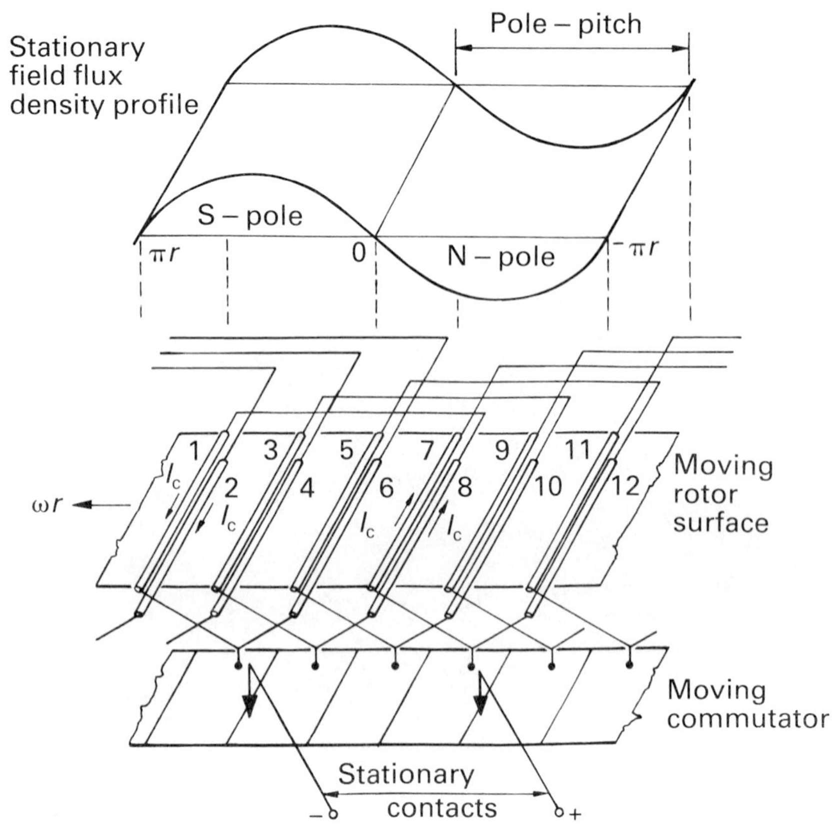

Fig. 7.4 The 12 conductor winding of Fig. 7.3 moved along by the distance between two adjacent slots.

where B_n is the flux density at the position of conductor n. As the conductors move B_n changes and e_n changes. Σe will likewise change until the rotor surface has moved on by an amount equal to the slot spacing, i.e. $2\pi/6$ radians in this case. Fig. 7.4 shows the new position of the conductors at this instant. Σe will now have the same value as it had for the position shown in Fig. 7.3 but the conductors contributing to Σe have changed in that conductors 2 and 7 have swapped paths with conductors 1 and 8. The basic periodicity for Σe is therefore the time it takes the rotor to travel $2\pi/6$ radians. At a constant speed of ω rad/sec this period is given by $\tau = 2\pi/6\omega$. The average value for Σe is therefore

$$E = \frac{1}{\tau}\int_0^\tau \Sigma e \,\mathrm{d}t = \frac{1}{\tau}\int_0^\tau (e_1 + e_2 + e_3 + e_4 + e_5 + e_6)\,\mathrm{d}t$$

$$= \frac{1}{\tau}\left[\int_0^\tau (e_1)\,\mathrm{d}t + \int_0^\tau (e_2)\,\mathrm{d}t \ldots + \int_0^\tau (e_6)\,\mathrm{d}t\right]$$

If we now change the variable on the RHS from t to θ by using the substitution

$$\mathrm{d}t = d\theta/\omega$$

and adjust the limits of integration accordingly then

$$E = \frac{1}{\omega\tau}\left[\int_{2\pi/3}^{\pi} e_1 lr\ \mathrm{d}\theta + \int_{2\pi/3}^{\pi} (e_2)\mathrm{d}\theta \ldots + \int_{0}^{\pi/3} (e_6)\mathrm{d}\theta\right]$$

$$= \frac{1}{\omega\tau}\left[\omega\left(\int_{2\pi/3}^{\pi} B_1 lr\ \mathrm{d}\theta + \int_{2\pi/3}^{\pi} B_2 lr\ \mathrm{d}\theta + \ldots + \int_{0}^{\pi/3} B_6 lr\ \mathrm{d}\theta\right)\right].$$

From Fig. 7.3 and Fig. 7.4 the individual integrals combine to give

$$E = \frac{2}{\tau}\left(\int_0^{\pi} Blr\ \mathrm{d}\theta\right) \tag{7.1}$$

This is precisely the same expression as we would get if we took the average e.m.f. for one conductor as it moved from $\theta = 0$ to $\theta = \pi$ and multiplied by the number of conductors in series, i.e.

$$E = 6 \times \frac{1}{3\omega\tau}\int_0^{\pi} \omega Blr\ \mathrm{d}\theta = \frac{2}{\tau}\int_0^{\pi} Blr\ \mathrm{d}\theta$$

Although this latter way of arriving at a value for E masks the basic periodicity in the total e.m.f. waveform it greatly facilitates the calculation of E in more complex windings and we shall make use of this viewpoint in later calculations. Returning to equation (7.1) the right-hand side can be simplified to give

$$E = \frac{2}{\tau} \times \phi$$

where ϕ is the integral of B over the rotor surface between the limits $\theta = 0$ and $\theta = \pi$. It is also in this case the field flux per pole Φ. Had the brush position been such that we were summing e.m.f's in conductors from $\theta = \alpha$ to $\theta = \pi + \alpha$, say, then ϕ would be less than Φ because the flux density changes sign in the range of integration.

Substituting for τ we get

$$E = \frac{12\omega}{2\pi} \times \Phi \text{ volts}$$

$$= 6 \times \Phi \Big/ \frac{\pi}{\omega} \text{ volts}$$

= number of conductors in series × (flux per pole ÷ time taken to traverse one pole 'pitch')

In general if there were N conductors in series between the brushes and

p poles (where N and p are both even) then

$$E = N \times \Phi \bigg/ \left(\frac{2\pi}{p\omega}\right) \text{ volts}$$

$$= Np\Phi\omega/2\pi \text{ volts} \tag{7.2}$$

The quantity $\Phi/(2\pi/p\omega)$ is the average e.m.f. in one conductor as it moves from one commutation position to the next, i.e. one pole-pitch.

When there are more than two poles there are two basic winding patterns which can be used to link up conductors.

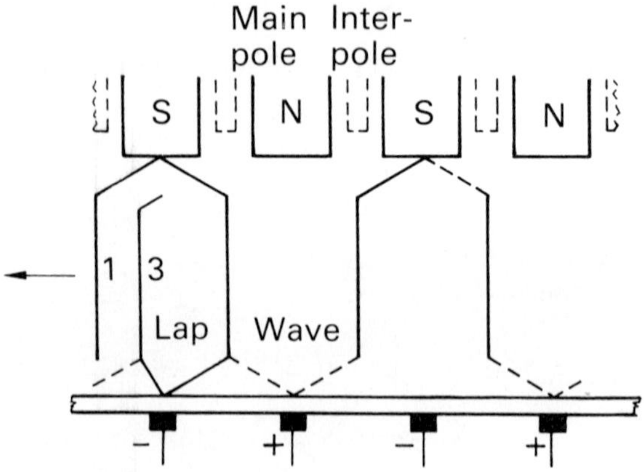

Fig. 7.5 Basic lap and wave winding patterns.

Fig. 7.5 shows the two types of pattern for a four-pole machine. The full line shows the type of pattern used in Figs 7.3 and 7.4 and this is called a LAP winding. Brushes of like polarity can be joined and bearing in mind that as shown in Fig. 4.12(b) the current splits into two paths at each positive or negative brush then there are four parallel paths through this winding. In general for each pole pair there will be two parallel paths and for a p pole machine 'a' $= p$.

The dotted line shows what is known as a WAVE winding whereby the pattern goes right round the armature before returning to conductor 3. Note that the two positive brushes are already joined by conductors lying in the zero field region. If the full wave winding pattern had been shown the negative brushes would also have been seen to be joined by other conductors in the zero field region. There is theoretically no need for more than one positive and one negative brush on a WAVE winding. The number of parallel paths through a wave winding is always two irrespective of the number of poles. In practice more than one pair of brushes is used in order to share the current pick up from the commutator.

LAP windings join conductors under successive pole-pairs in parallel while WAVE windings join them in series.

In equation (7.2) the quantity N will therefore depend on the type of winding. This can be catered for quite simply by expressing N in terms of the total number of conductors (coil sides) Z and the number of parallel paths, 'a', i.e.

$$N = Z/a \tag{7.3}$$

$$\therefore \quad E = p\Phi Z\omega/2\pi a \tag{7.4}$$

For a given machine p, Z and 'a' will be fixed so that we may write

$$E = K\Phi\omega \tag{7.5}$$

where

$$K = \frac{pZ}{2\pi a} \tag{7.6}$$

Torque may be calculated in a similar fashion to the e.m.f. either by considering the average value of instantaneous torque over the period τ or by considering the average torque on one conductor as it moves over a pole-pitch and then multiplying by the number of conductors. The latter technique as we saw with the e.m.f. calculation is much simpler.

If we consider conductor 6 in Fig. 7.4 then the instantaneous torque on this conductor is

$$T_c = B_6 l I_c r$$

I_c is the conductor current and is constant as conductor 6 moves from $\theta = 0$ to $\theta = \pi$.

$$\therefore \quad \frac{\omega}{\pi}\int_0^{\pi/\omega} T_c \, dt = \frac{\omega}{\pi}\int_0^{\pi/\omega} BlrI_c \, dt$$

$$= \frac{I_c}{\pi}\int_0^{\pi} Blr \, d\theta$$

$$= \frac{I_c\Phi}{\pi} \tag{7.7}$$

In the case where there are p poles then the angular span of one pole-pitch is $2\pi/p$ radians and the average torque for one conductor is

$$\omega/(2\pi/p)\int_0^{2\pi/p\omega} T_c \, dt = \frac{p}{2\pi}\int_0^{2\pi/\omega p} (Blr \, d\theta)I_c$$

$$= \frac{pI_c\Phi}{2\pi} \tag{7.8}$$

(Note that the integral of field flux density over the pole-pitch is always Φ.)

If there are now Z conductors with 'a' parallel paths through the armature then the total average torque T is

$$T = \frac{pZ\Phi I}{2\pi a} \tag{7.9}$$

where $I_c = I/a$ and I is the total d.c. current into or out of the machine.

For a given machine equation (7.9) can be stated as

$$T = K\Phi I \tag{7.10}$$

where K is as defined in equation (7.6).

In an actual machine where Z is much larger than the 12 in our simple winding, the ripple in T and E is very small. The field flux density profile is also much closer to a square wave than to the sine wave we have considered so far and this too reduces the ripple due to interconductor spacing. On the other hand, practical machines have conductors embedded in slots and this produces a 'tooth' ripple. The commutation process may also vary the number of conductors in series between brushes so that for one reason or another the e.m.f. or torque will have a small ripple.

The reader may satisfy himself at this stage that $\omega T = EI$.

7.4 COMMUTATION

Students invariably find the commutation process difficult to understand and since it does not enter directly into the equations of performance of the machine it could be omitted in an elementary text such as this. On the other hand it is the commutator which limits the size of d.c. machines, demands most maintenance and greatly contributes to machine costs. Such an important process demands a little more attention than we have given it so far.

In Fig. 7.3 the width of the brush was reduced to facilitate the calculation of e.m.f. Quite apart from the problem of current density at a point contact there is another important defect brought into play by limiting the brush width. The arrows on the conductors in Figs 7.3 and 7.4 indicate that the current direction is reversed when a coil side changes from one group to another. The coil sides in question are 1, 2, 7 and 8. This change takes place as the brush moves from one commutator segment to the next and for a point contact would be instantaneous. The self-inductance of the coils in which the current is changing will give rise to high $L\,di/dt$ voltages in these coils which in turn would lead to arcing between the brush and the commutator segment it had just left. This leads to rapid deterioration of the commutator. A finite time must be allowed for the change of current direction in coil sides passing from one group to another.

Fig. 7.6(a) shows the 12 conductor armature with 2 conductors per slot

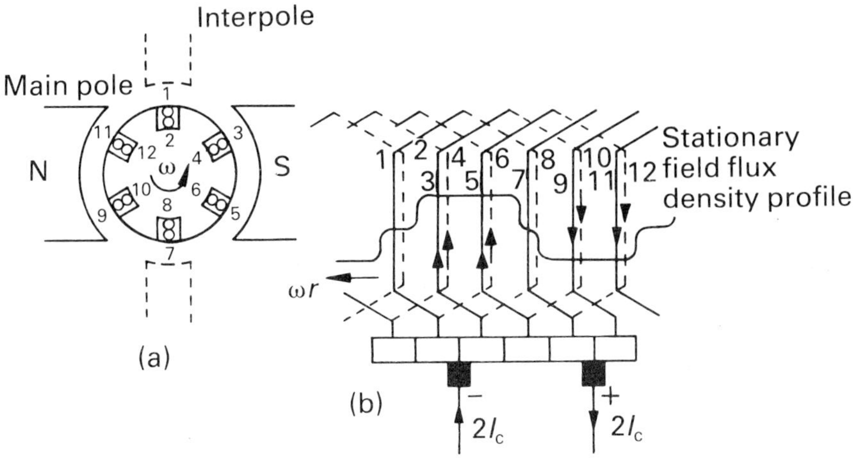

Fig. 7.6 (a) 12 conductor slotted armature and (b) development of (a) showing coils shorted by brushes.

and Fig. 7.6(b) shows a development of the armature winding for the rotor position shown in Fig. 7.6(a). The brushes have been given a finite width and the field flux density distribution is approximately a square wave. From Fig. 7.6(b) the negative brush is closing a loop through coil sides 1 and 8 and the positive brush is closing a loop through coil sides 2 and 7. The arrows on the coil sides indicate the direction of e.m.f. and current. Coil sides 1, 2, 7 and 8 will be carrying current but they have no e.m.f. arising from cutting field flux. The current in these coil sides is a circulating current and differs from the normal conductor current I_c. Let us look specifically at the

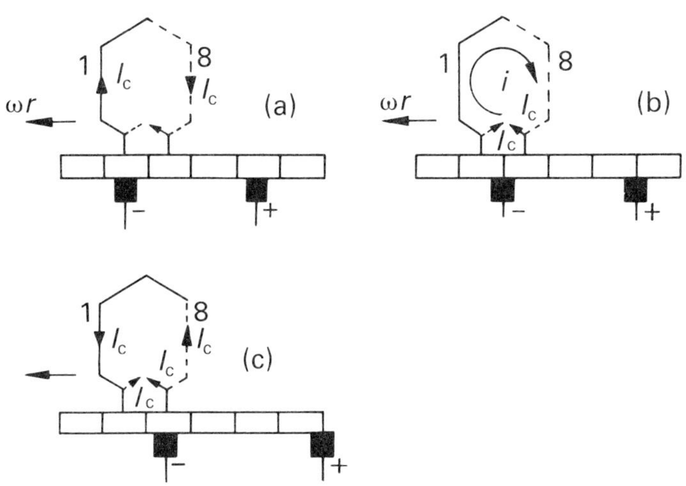

Fig. 7.7 Commutation of currents in coil sides 1 and 8. (a) Prior to commutation, (b) during commutation and (c) after commutation.

Fig. 7.8 Main poles and interpoles on the magnet frame of a 8.2 MW, 76 r.p.m. mine winder drive d.c. motor. Note the heavy bar winding on the interpoles to carry the armature current. (Courtesy of G.E.C. Large Machines Ltd).

coil comprising conductors 1 and 8 at three instants of time. Fig. 7.7(a) shows an instant prior to the position shown in Fig. 7.6(b) and corresponds to Fig. 7.3. Because the brush is in the middle of a commutator segment no coils are shorted and all coil sides contribute an e.m.f. Fig. 7.7(b) is the instant corresponding to Fig. 7.6 when a loop exists via the negative brush and Fig. 7.7(c) is the instant corresponding to Fig. 7.4 when all conductors again contribute an e.m.f. but both the e.m.f. and current in coil sides 1 and 8 have reversed in comparison to Fig. 7.7(a). The same process is taking place at the positive brush involving coil sides 2 and 7. During the period when the negative brush short-circuits coil sides 1 and 8 the circulating current i shown in Fig. 7.7(b) must go from I_c in a clockwise direction to I_c in an anticlockwise direction. The self-inductance of the loop will oppose this change and furthermore the flux density set up by the currents in the other armature conductor (armature reaction, Section 4.7.1) at the location of conductors 1 and 8 is such as to generate an e.m.f. also opposing the change. In order to produce an e.m.f. in coil sides 1 and 8 which will force a reversal of i during the commutation period an additional 'field' flux is required in the space between the main poles where the coil sides undergoing commutation are situated. This additional field is set up by small poles called interpoles or compoles and these are shown dotted in Fig. 7.6(a). The interpoles are excited by a winding carrying the armature output current ($2I_c$) and the strength of the magnetic field under the interpoles therefore varies with the magnitude of the armature current.

Even with interpoles ideal commutation is never achieved over the complete range of load currents and a limited amount of sparking, especially at the trailing edge of a brush, is inevitable.

Fig. 7.8 shows a d.c. machine in which the main pole and interpole structure can be distinguished.

While coil sides are shorted out by one or other of the brushes they clearly do not contribute to the e.m.f. between brushes. For instance there are only 4 coil sides in each group contributing to the e.m.f. for the armature position of Fig. 7.6(b) compared to 6 coil sides in Figs 7.6(a) and (c). In machines with a small number of slots per pole this can produce a significant voltage ripple but in machines with a large number of coil sides per pole the loss of a few coil sides during commutation is negligible.

7.5 ARMATURE REACTION

In Chapter 4 we saw that armature reaction was an essential feature in the torque production process. It does, however, have undesirable 'side effects' in the d.c. machine. Fig. 7.9(a) shows the field flux acting alone and Fig. 7.9(b) shows the flux produced by armature currents acting alone. Note that

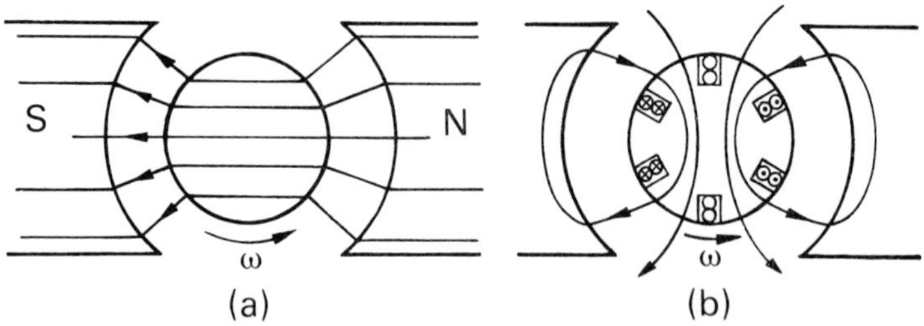

Fig. 7.9 (a) Flux produced by main pole winding and (b) flux produced by armature currents.

in the generator case shown here the field flux and armature flux oppose each other at the leading pole tips and aid each other at the lagging pole tips. Were the magnetic path material linear then Φ, the flux per pole, would remain unchanged since the effects of armature flux would cancel out over the pole pitch. Due to saturation, however, the increase in resultant flux density at the lagging pole tips is less than the decrease at the leading pole tips with a resultant decrease in Φ. (We can add m.m.f. but not flux density in a non-linear magnetic medium.)

The other effect to note is that there is now a non-zero flux density over the two slots containing the conductors undergoing commutation and the direction of this flux is such as to hinder rather than help commutation.

This latter effect of armature reaction is overcome by the use of interpoles as we mentioned in Section 7.4 and the former effect can be overcome by the use of a compensating winding in the pole-face. The vast majority of machines make do without a compensating winding since such windings are very expensive and can only be justified in special cases.

7.6 METHODS OF EXCITATION

Fig. 7.8 shows that each main pole carries a winding and the main pole flux (Φ) is controlled by the current in this winding.

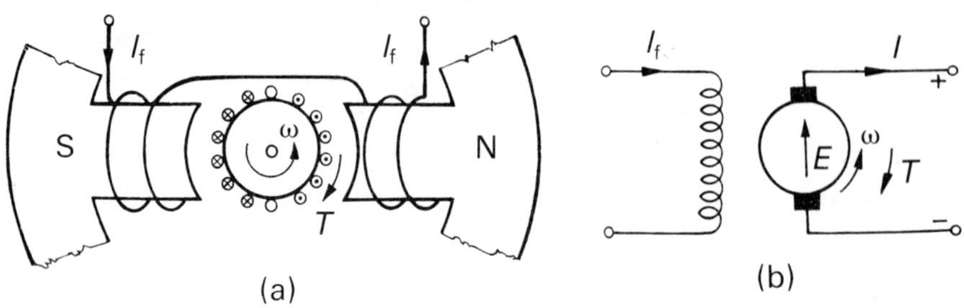

Fig. 7.10 (a) Generator, showing direction of field and armature currents and (b) circuit diagram for (a).

Fig. 7.10(a) shows a diagrammatic representation of a two-pole generator and Fig. 7.10(b) shows how the field and armature circuits are represented in a circuit diagram. The directions of E and I in Fig. 7.10 are consistent with generator action. The armature has to be driven against a counter torque to produce an electrical output. Reversal of current direction in both diagrams will give T in the same direction as ω, i.e. a mechanical output resulting from a corresponding electrical input.

The current I_f can be derived in several ways:

(1) By supplying the field winding from an independent source.
(2) By connecting the field winding in parallel with the armature.
(3) By connecting the field winding in series with the armature.

Method (1) is known as 'separate excitation', method (2) as 'shunt excitation' and method (3) as 'series excitation'. By having two separate windings on the poles it is possible to mix methods (2) and (3). This technique is called 'compound excitation' with one winding shunt connected and the other series connected. The interconnection between the field and armature windings greatly affects the output characteristics of the machine. In small machines up to approximately 1 kW rating the field may be supplied by a permanent magnet incorporated into the pole structure. The flux per pole cannot be varied in a permanent magnet machine and certain types of permanent magnet material which are susceptible to being demagnetised by armature reaction m.m.f. require the use of a compensating winding in the pole face. (2)

7.7 LOSSES IN D.C. MACHINES

There are the obvious I^2R losses in field and armature circuit resistances. The brush resistance also contributes an I^2R loss but the R in this case is non-linear due to the characteristics of the normal brush material carbon graphite. The brush resistance drops as the current increases such that the IR product remains roughly constant. In the absence of any more exact information we shall assume that there is a 1 volt drop across a brush.

Iron losses occur in the rotor since it experiences flux changes as it rotates through the stationary field flux. Small iron losses also occur in the pole faces due to flux ripples set up by the passage of the rotor teeth. Despite the latter the poles and yoke on which they are mounted are normally made from solid iron whereas the rotor is assembled from laminations (called stampings). Special machines where rapid changes of Φ are required may also have laminated poles and yoke. The iron losses are a 'rotational' loss.

The other rotational losses are windage and friction.

In a load test when all of the above-mentioned losses have been taken into account the actual efficiency turns out to be slightly lower than calculated. There is a further loss only present when the machine is loaded which carries the name 'stray load loss'. It arises from a number of sources such as increased rotor iron losses due to leakage flux around armature conductors and skin effects in armature conductors. (Remember that the armature conductors have the current direction reversed as they go through commutation and the d.c. resistance of the armature will therefore be less than the resistance met by load currents.) Stray load loss can usually be taken to be 1 per cent of output unless more detailed information is available.

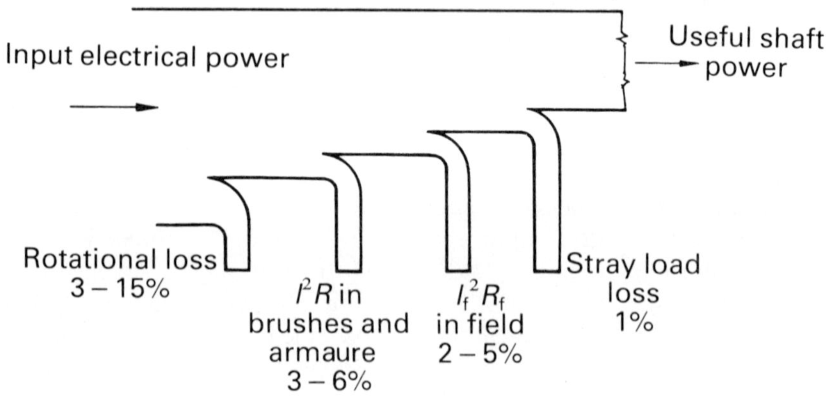

Fig. 7.11 Losses in a d.c. motor.

Fig. 7.11 shows a typical power breakdown for a d.c. motor. The lower percentage losses tend to occur in large machines.

7.8 MACHINE CHARACTERISTICS

The relationship between the flux per pole Φ and ampere-turns on the field winding is a fundamental characteristic of any machine. Fig. 7.12(a) shows the typical shape of this characteristic. It does not go through zero due to the residual magnetism of the iron and is not linear for high IN due to saturation. The flux per pole Φ and field ampere-turns are not variables which can be readily measured but the shape of this curve is reflected in the relationship between armature e.m.f., E and field current I_f at a fixed speed ω. I_f is supplied from a separate source for the purpose of deriving this curve. Equation (7.5) tells us that E is proportional to Φ for a fixed ω and is proportional to ω for a fixed Φ. The curves in Fig. 7.12(b) are alternatively known as the saturation characteristic or the open-circuit characteristic (O.C.C.). The latter is the more common.

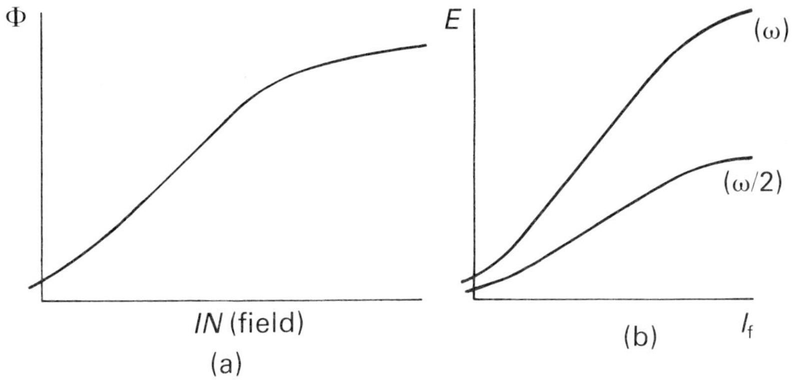

Fig. 7.12 (a) Magnetisation characteristic and (b) open-circuit characteristic at two speeds.

7.9 THE SHUNT-CONNECTED MOTOR

Fig. 7.13(a) shows how the field and armature circuits are interconnected. An external resistor R_f is connected in series with the field. R_f is normally a rheostat and is used to control the field current. The current directions

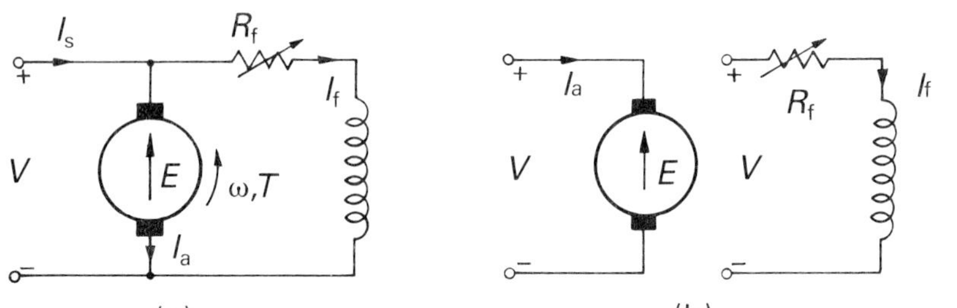

Fig. 7.13 Circuit diagrams for (a) shunt-connected motor and (b) separately excited motor.

shown are for a motor connected to a supply V drawing a supply current I_s. To distinguish between armature and field currents the armature current is labelled I_a and the field current I_f. There is no difference in the motor case between the connections shown in Fig. 7.13(a) and (b), i.e. between shunt excitation or separate excitation. I_f in both cases is independent of E and dependent only on V and R_f. The characteristics of a shunt or separately excited motor are therefore the same.

The output characteristic for a motor is the speed/torque curve. The equations available to derive this relationship are as follows:

$$E = K\Phi\omega \qquad (7.5)$$

$$T = K\Phi I_a \tag{7.10}$$

The O.C.C.,

$$E_{(\omega_1)} = \text{fn}(I_f) \tag{7.11}$$

$$I_f = V/R_f \tag{7.12}$$

$$V = E + I_a R_a + 2 \tag{7.13}$$

In these equations R_f is taken to include the field winding resistance, R_a is the resistance of the armature current path excluding the brushes but including interpole windings when present and the 2 in equation (7.13) is the brush volt drop. The O.C.C. is a plot of E versus I_f at a constant speed of ω_1 rad/s. It can be put in a much more useful form for dealing with motor application if we divide the e.m.f. values by ω_1 and use equation (7.5), i.e. $E_{(\omega_1)}/\omega_1 = K\Phi = E/\omega$ to get a plot of $K\Phi$ against I_f. Let us call this modified plot equation (7.14). (7.14)

For a particular setting of R_f and a constant supply voltage, V, I_f can be determined from equation (7.12). This value of I_f will then determine the value of $K\Phi$ from the modified O.C.C. of equation (7.14). I_a can now be found from equation (7.10) when a particular value of gross torque T is specified. This value of I_a inserted in equation (7.13) determines E and since $K\Phi$ is already known we can now find ω from equation (7.5). Therefore, having set I_f and specified T we can determine ω. How much of the gross torque is available as shaft output torque will depend on the 'lost torque' used to overcome the rotational losses at the speed concerned.

The general shape of the torque speed curve can be determined by substituting from equations (7.5) and (7.10) into equation (7.13) to get

$$V = K\Phi\omega + \frac{TR_a}{K\Phi} + 2$$

$$\therefore \quad \omega = \frac{(V-2)}{K\Phi} - T\left(\frac{R_a}{(K\Phi)^2}\right) \tag{7.15}$$

Equation (7.15) gives a straight line characteristic when $K\Phi$ is fixed and provided armature reaction effects are small $K\Phi$ is only dependent on the value of I_f. The slope $R_a/(K\Phi)^2$ is quite small resulting in speed changes from no load to full load torque being of the order of 5 per cent. Fig. 7.14(a) shows a set of speed torque curves for various values of R_f. The minimum value which R_f can take is the winding resistance of the field. Since shunt field windings have a large number of turns the minimum value is likely to be of the order of 100 Ω.

The dotted line on Fig. 7.14(a) shows that as I_f and hence $K\Phi$ decrease

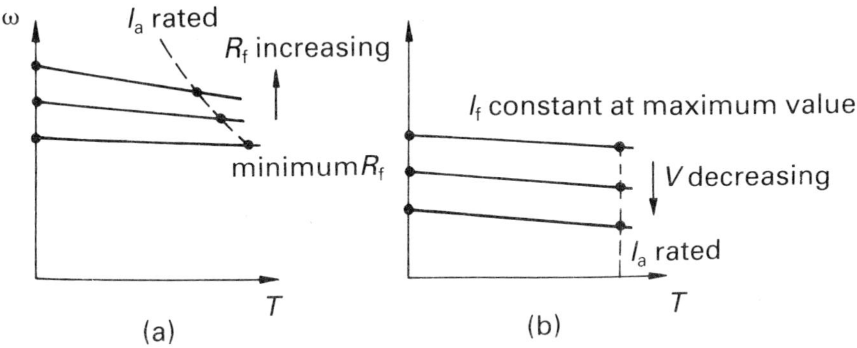

Fig. 7.14 (a) Speed/torque curves at constant V, R_f varying and (b) speed/torque curves at constant I_f, V varying.

then the torque corresponding to the rated value of I_a will decrease (equation (7.10)). If armature reaction effects are taken into account then $K\Phi$ decreases slightly as I_a increases, i.e. as T increases. The net result is that because of the predominance of the first term in equation (7.15) ω is slightly higher than for the corresponding case with no armature reaction effects. As I_f decreases the proportional effect of armature reaction in weakening Φ increases such that ω may rise as I_a increases which may lead to instability. For this reason I_f is rarely decreased to less than 70 per cent of its maximum value with the field rheostat set to zero. Where a wider speed range is required using I_f control the machine must either be fitted with a compensating winding or a small amount of positive feedback from the armature current to the flux per pole Φ can be employed. The latter is more common and will be discussed in Section 7.11. In either case I_f can then be reduced by a factor of about 3 and a useful torque obtained before the rated current is exceeded.

When the speed has to be reduced beyond the value obtained with I_f at its maximum value then V in equation (7.15) can be reduced while $K\Phi$ remains at its maximum value. This requires that the field circuit be supplied from a separate source which remains at the original voltage while the voltage applied to the armature circuit, V, is reduced. Fig. 7.14(b) shows the speed/torque curves obtained in this way. Note that the rated current limit is now at a constant torque value (why ?). Armature voltage control can be used to give stable speeds down to zero with the torque depending only on the value of armature current.

Example 7.1

A 400 V shunt-connected d.c. motor takes a total current of 3.5 A on no load and 59.5 A on full load. The field circuit resistance is 267 Ω and the armature circuit resistance is 0.2 Ω (excluding brushes). If the armature reaction effect at 'full load' weakens the flux per pole by 2 per cent compared to its no load value find the percentage change in speed from no load to full load. What would be the change in speed if the flux per pole remained constant?

$$I_f = 400/267 = 1.5\,\text{A}$$

$$\therefore \quad I_a \text{ on no load} = 2\,\text{A}, \; I_a \text{ on full load} = 58\,\text{A}$$

$$E \text{ on no load} = 400-2-2 \times 0.2 = 397.6\,\text{V}$$

$$E \text{ on full load} = 400-2-58 \times 0.2 = 386.4\,\text{V}$$

From equation (7.5)

$$E = K\Phi\omega$$

$$\therefore \quad \text{on no load} \qquad 397.6 = K\Phi\omega_1$$

$$\text{on full load} \qquad 386.4 = K(0.98\Phi)\omega_2$$

$$\therefore \quad \frac{\omega_2}{\omega_1} = \frac{386.4}{397.6} \times \frac{1}{0.98} = \frac{0.97}{0.98} = \underline{0.99}$$

There is a 1 per cent drop in speed. When there is no flux weakening the 0.98 factor disappears from the ratio and there is 3 per cent drop in speed.

This example shows that the basic reason for the drop in speed is the I_aR_a voltage leading to a decrease in E. Because this drop is small even at full load the intrinsic speed drop is low and only a small amount of flux weakening is required to counteract the effect. The actual speed drop would probably be higher in the above example due to the brush drop at 2 A being less than 2 V.

Example 7.2

What resistance must be placed in series with the armature in the above machine if the full load speed is to be reduced by 50 per cent with the gross torque remaining constant? R_f remains at 267 Ω.

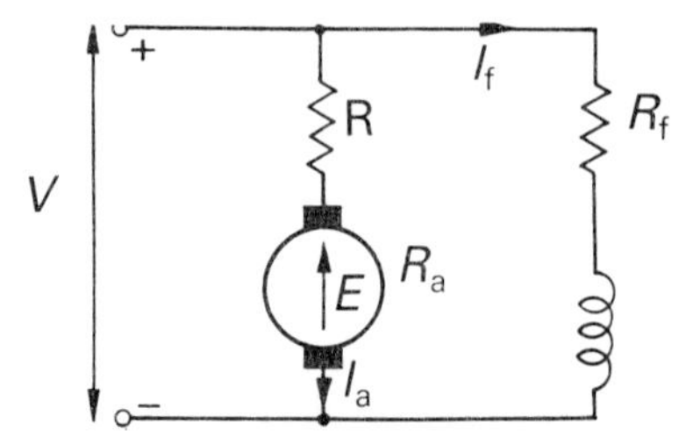

Fig. 7.15 Circuit diagram for Example 7.2.

From equation (7.10),

$$T = K\Phi I_a$$

This tells us that provided Φ is constant (or $0.98\,\Phi$ in the first case) the gross torque will remain constant if I_a remains constant

$$\therefore \quad I_a = 58\,\text{A}$$

With no R in the armature circuit of Fig. 7.15

$$E = 400 - 2 - 58 \times 0.2 = 396.4\,\text{V}$$

and from the equation (7.5) if ω is to be reduced by 50 per cent while the flux per pole remains constant then E must also reduce by 50 per cent.

$$E \text{ in Fig. 7.15} = \frac{386.4}{2} = 193.2\,\text{V}$$

$$\therefore \quad 400 = 193.2 + 2 + 58R + 58 \times 0.2$$

$$58R = 400 - 193.2 - 2 - 11.6 = 193.2\,\text{V}$$

$$\underline{R = 3.33\,\Omega}$$

This is clearly an inefficient method of controlling speed because of the power wasted in R.

7.10 THE SHUNT-CONNECTED GENERATOR

In comparison to the motor circuit diagram of Fig. 7.13(a), the armature current I_a in Fig. 7.16(a) is in the same direction as E indicating that E is now the source of power. ω and T are in opposite directions indicating that the generator requires a mechanical input to drive the armature round

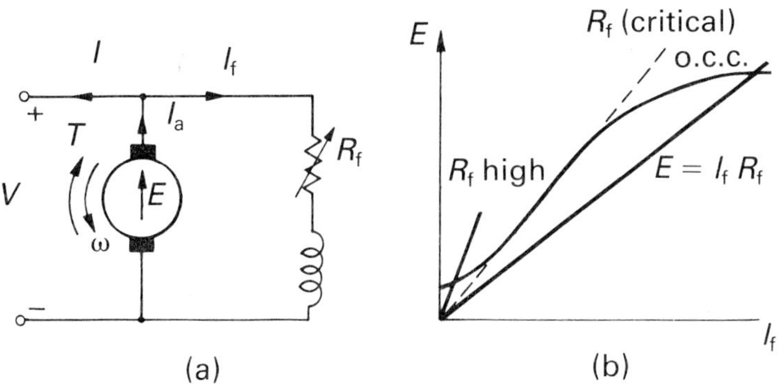

Fig. 7.16 (a) Shunt-connected generator and (b) solution for E when $I = 0$.

against the counter torque T. The voltage V may be a fixed supply into which the generator is delivering power or it may simply be the voltage across a load resistor R_L say. In the former case the machine behaves like a separately excited generator with I_f decided by the setting of R_f and the supply voltage. The generator must then be driven fast enough for E to exceed V in order that power flow into the supply. If for any reason ω falls below the value at which $E = V$ then the machine will draw power from the supply and deliver mechanical power to the shaft, i.e. behave like a motor. If the shaft speed is fixed then the mode of operation, motoring or generating, can be altered by altering R_f to adjust the magnitude of E relative to V.

When V is the voltage across a load resistor then V is no longer independent of I_f. Since I_f depends on V and R_f, finding V and I_f for a particular setting of R_L is not a straightforward process.

Let us consider the simple case of R_L set at infinity, i.e. the generator is on open circuit. With I set to zero in Fig. 7.16(a) then

$$E = I_f(R_a + R_f) \tag{7.16}$$

$$E = \text{fn}(I_f) \tag{7.17}$$

Equation (7.17) is the O.C.C. plotted for the speed at which the generator is rotating with I_f supplied from a separate source. In equation (7.16) $R_a \ll R_f$ and because $I_f \ll I_a$ (rated) the brush drop is omitted. Equation (7.16) can therefore be modified to

$$E = I_f R_f \tag{7.18}$$

Fig. 7.16(b) shows a plot of $E = \text{fn}(I_f)$ and $E = I_f R_f$ for several values of R_f. The intersection of the O.C.C. and the line representing $I_f R_f$ gives the solution of equations (7.17) and (7.18). For high values of R_f the voltage is quite low and the machine would be said to 'fail to self-excite'. As R_f decreases there comes a value of R_f whereby the line $I_f R_f$ is nearly colinear with the O.C.C. This value of R_f marks the transition between a low value of E and a high value of E and is called the critical resistance for the particular speed for which the O.C.C. is plotted. Further decreasing R_f leads to high values of E at which the machine would be said to 'self-excite'. The critical field resistance will change as ω changes with R_f (critical) increasing as ω increases.

When R_L is finite then the circuit equations become

$$E = \text{fn}(I_f) \tag{7.17}$$

$$V = E - I_a R_a - V_b \tag{7.19}$$

$$V = I_f R_f = IR_L \tag{7.20}$$

$$I = I_a - I_f \tag{7.21}$$

V_b is the brush voltage which will be small for low values of I_a, rapidly increasing to about 2 V as I_a increases and staying roughly constant thereafter.

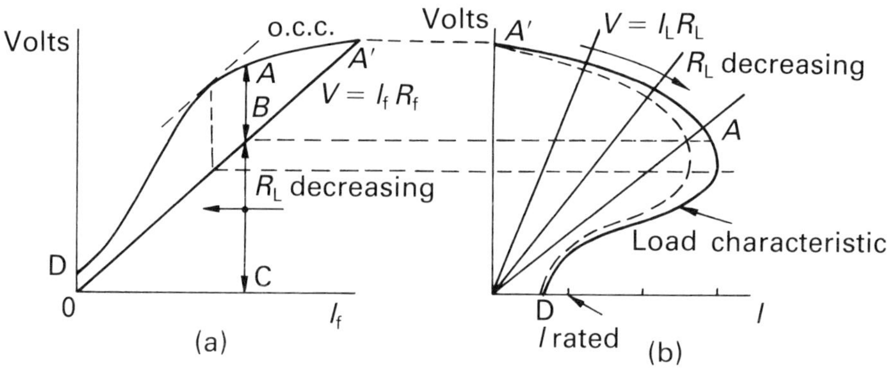

Fig. 7.17 (a) Derivation of typical points on the load characteristic and (b) the load characteristic.

If we consider Fig. 7.17(a) and start with R_L equal to infinity then A′ gives the relevant values for V, E and I_f, as in the open-circuit case. As R_L is decreased V will differ from E due to the various volt drops given in equation (7.19) resulting from an increase in I_a. This decrease in V will cause I_f to drop and this in turn leads to a decrease in E. The operating point on the O.C.C. in Fig. 7.17(a) will move from A′ towards a typical point A. At A the vertical line AC represents E, 0C represents I_f and BC represents $V = I_f R_f$. The remaining portion AB is the difference between E and V when I_f has the value 0C. From equation (7.19)

$$\text{AB} = E - V = I_a R_a + V_b$$

$$\therefore \quad I_a = \frac{\text{AB} - V_b}{R_a} \tag{7.22}$$

Except for very low values of I_a when AB ≤ 2 volts we can put $V_b = 2$ into equation (7.22) and obtain a value for I_a. We can therefore plot the generator load characteristic by sweeping a vertical line on Fig. 7.17(a) from A′, through A to D and at each position break it down into components AB and BC. BC gives V directly and AB gives I_a from equation (7.22). The variation in I_a is roughly proportional to AB and $I = (I_a - I_f)$ is likewise roughly proportional to AB due to I_f being small in comparison to I_a except for points close to A′. As R_L decreases the line sweeps from right to left on Fig. 7.17(a) and the load characteristic is as shown in Fig. 7.17(b). Points A′, A and D are shown on this characteristic and the value of I for a particular value of R_L can be found from the intersections of the load characteristic and the $V = IR_L$ line of appropriate slope. Note that the maximum current is not when the machine terminals are short-circuited. A typical ratio of rated to maximum current is shown on Fig. 7.17(b).

Throughout the above process R_f and ω have been assumed constant and the effects of armature reaction in weakening the flux per pole have been neglected. The latter effect will decrease I for a given V on the machine load characteristic but will not change the basic shape of this characteristic. The dotted curve on Fig. 7.17(b) shows the slight change in load characteristic incorporating armature reaction effects.

Example 7.3

The open-circuit characteristic for a d.c. generator running at 1000 rev/min is:

Field current (A)	0	2	4	6	8
Armature volts (V)	10	90	170	214	240

Estimate the critical field resistance when the machine is on open-circuit and run at 1000 rev/min. With the field resistance set at this value the machine is run at 1500 rev/min. Estimate the output voltage on no load.

The machine is connected to a load resistor such that the armature current is 400 A. If the armature circuit resistance other than brushes is 0.063 Ω estimate the output voltage and load current at 1500 rev/min. Neglect armature reaction effects.

We must first of all plot the O.C.C. at 1000 rev/min and 1500 rev/min. The latter is derived by increasing all the armature voltage values by 50 per cent since for a given I_f (and hence Φ) E is proportional to ω.

Fig. 7.18 shows these curves with the 1000 rev/min curve dotted.

The $I_f R_f$ line shown is co-linear with the linear portion of the 1000 rev/min O.C.C. and this line passes through 174 volts at 4 A.

$$\therefore \quad R_f \text{ critical} = 174/4 = \underline{43.5\ \Omega}$$

At 1500 rev/min the above field resistance is no longer the critical value and the intersection of the $I_f R_f$ line and the O.C.C. for 1500 rev/min gives an output voltage of approximately $\underline{364}$ volts.

When the generator is loaded the difference (see equation (7.19))

$$\begin{aligned}(E - V) &= I_a R_a + 2 \\ &= 400 \times 0.063 + 2 \\ &= 27.2\ \text{V}\end{aligned}$$

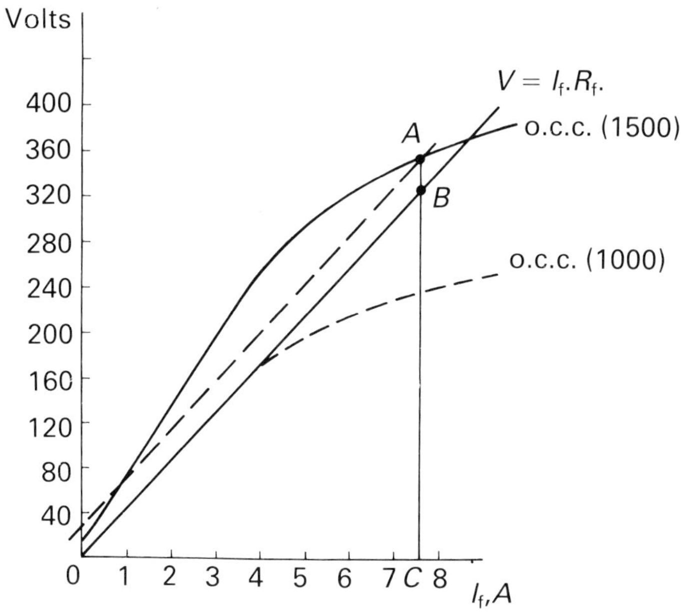

Fig. 7.18 Graphical construction for Example 7.3.

A vertical line ABC can now be moved along Fig. 7.18 to find a position at which AB = 27.2 V. A simple technique to find the position is to draw a line parallel to the $I_f R_f$ line but 27.2 V above it. This will locate the point A.

For the position shown

$$V = \text{BC} \doteqdot \underline{328\ \text{V}}$$

$$I_f \doteqdot 7.6\ \text{A}$$

$$\therefore \quad I = I_a - I_f = \underline{392.4\ \text{A}}$$

7.11 THE SERIES-CONNECTED MOTOR

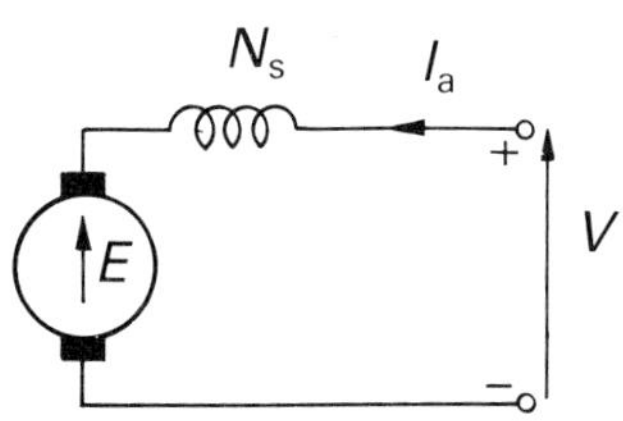

Fig. 7.19 The series connected motor.

The field excitation is now derived by passing the armature current through the field winding on the poles. The armature current is much larger than the field current in the shunt-excited machine and the number of turns N_s in the field winding is much less than in the shunt case. The product field current × turns in the field winding is similar to that for the shunt case in order to give similar values for Φ. The series field winding resistance is very much less than the shunt field resistance due in part to the smaller number of turns and in part to the much larger cross-section of the conductor in the series field.

The torque speed characteristic can be derived from equation (7.5), (7.10), the magnetisation characteristic and the circuit equation of Fig. 7.19. That is

$$E = K\Phi\omega \tag{7.5}$$

$$T = K\Phi I_a \tag{7.10}$$

$$\Phi = \text{fn (field current)} = \text{fn}(I_a) \tag{7.23}$$

and

$$V = I_a(R_a + R_s) + 2 + E \tag{7.24}$$

$$\therefore \quad V = \frac{T}{K\Phi}(R_a + R_s) + 2 + K\Phi\omega \tag{7.25}$$

Equation (7.25) contains $K\Phi$ which is a function of the armature current I_a via the magnetisation characteristic. An exact relationship between T and ω depends on having an exact analytical expression relating Φ to I_a. We could fit a polynomial to the measured magnetisation characteristic in order to achieve this exact relationship but if we will settle for an approximate torque–speed curve there is no need for such complication.

There are two regions in which the magnetisation characteristic is a simple function. For low values of I_a the relationship is linear and for very high values of I_a, Φ tends to a constant value. We can use these relationships to describe the two extremes of the torque–speed curve and then infer a smooth transition from one region to the other. In the linear region current and flux, and hence torque, are small and in the saturated region current, flux and hence torque are large.

Over the linear region of the magnetisation characteristic we can say

$$K_1\Phi = I_a \tag{7.26}$$

Substituting this relationship into equation (7.10) gives

$$T = KK_1\Phi^2$$

$$\therefore \quad \sqrt{T} = \sqrt{(KK_1)}\,\Phi$$

and

$$\therefore \quad K\Phi = \sqrt{\frac{K}{K_1}}\,\sqrt{T} = k\sqrt{T} \tag{7.27}$$

Substituting this relationship into equation (7.25) gives

$$V - 2 = \frac{\sqrt{T}}{k}\,(R_a + R_s) + k\sqrt{T}\omega$$

$$\omega = \frac{(V-2)}{k\sqrt{T}} - \frac{(R_a + R_s)}{k^2} \tag{7.28}$$

The second term in equation (7.28) arises from the term $I_a(R_a + R_s)$ in equation (7.24) and for the range of currents for which (7.26) is true this term is negligible in comparison to E.

$$\therefore \quad \omega \simeq \frac{(V-2)}{k\sqrt{T}} \qquad \text{i.e. } \omega^2 = \left(\frac{V-2}{k}\right)^2 \frac{1}{T} \tag{7.29}$$

This portion of the speed–torque curve is labelled ① in Fig. 7.20(a). In the saturated region we can say

$$K\Phi = k', \tag{7.30}$$

and this when substituted into equation (7.25) gives

$$V - 2 = \frac{T}{k'}\,(R_a + R_s) + k'\omega$$

$$\therefore \quad \omega = (V-2)/k' - T(R_a + R_s)/(k')^2 \tag{7.31}$$

This is a straight line relationship and is labelled ② in Fig. 7.19(a).

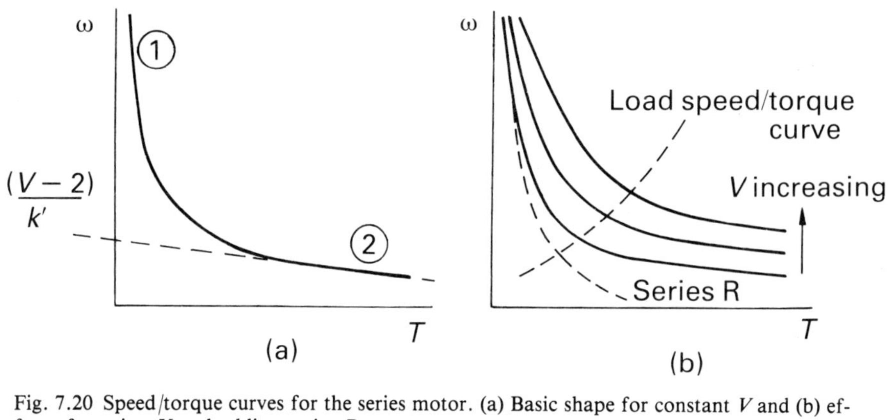

Fig. 7.20 Speed/torque curves for the series motor. (a) Basic shape for constant V and (b) effect of varying V and adding series R.

Unlike the shunt or separately excited motor speed falls rapidly as torque increases. The product ωT therefore varies much less and this type of speed–torque curve is ideal for traction purposes. It demands small variation in power over a wide range of torque values.

Fig. 7.20(b) shows how a set of torque–speed curves can be achieved by varying V. The speed at which a particular load would run will depend on the intersection of the load torque–speed curve and the motor torque–speed curve. The speed–torque curve can also be varied by inserting series resistance to increase R_s and this effect is shown by the dotted curve. The latter curve is for the same voltage V as the bottom full line curve on Fig. 7.20(b).

The solution to any d.c. machine problem requires the use of the magnetisation characteristic or some curve such as the O.C.C. which is dependent on the magnetisation characteristic. In the series motor the relevant curve is a plot of E/ω versus I_a. The plot is derived from a set of experimental results as in the following worked example. Flux weakening due to armature reaction effects is automatically taken into account as a result of I_a being present when $E/\omega(\propto\Phi)$ is calculated.

Example 7.4

A series motor has armature circuit resistance $(R_a + R_s)$ of 0.3 Ω. Test results when connected to a supply voltage of 1452 V were:

Current I_a(A)	100	150	200	250	300
Torque $T(10^3\,\text{Nm})$	37	77	125	165	207
Speed ω(rad/s)	3.64	2.60	2.18	1.92	1.80

Find the speed when the load torque is 150 000 Nm and the supply voltage is 1500 V. What does the speed drop to when a 3 Ω resistor is inserted in series with the armature? Assume that for a given current the lost torque is independent of speed.

From equation (7.10) the total torque T is

$$T = K\Phi I_a = KI_a\text{fn}(I_a) \tag{7.32}$$

where $\Phi = \text{fn}(I_a)$ is the magnetisation characteristic. The load torque T_L is

$$T_L = T - T_e \tag{7.33}$$

where T_e is the lost torque. Now since T_e is only a function of I_a and equation (7.32) shows that T is only a function of I_a then from equation (7.33), T_e is only a function of I_a. Therefore from a plot of T_L against I_a we can find I_a corresponding to 150×10^3 Nm. Fig. 7.21(a) shows this plot from which $I_a = 230$ A at 150×10^3 Nm.

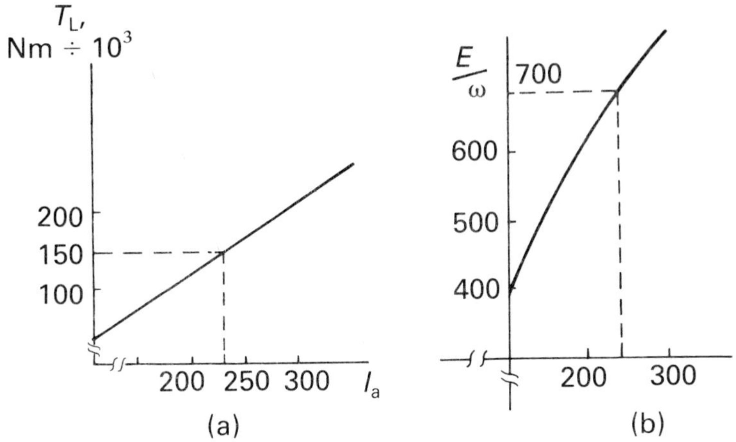

Fig. 7.21 Graphical construction for Example 7.4.

From Fig. 7.19 the circuit equation is

$$V = E + I_a(R_a + R_s) + 2$$

$$\therefore \quad 1500 = E + 230 \times 0.3 + 2$$

$$\therefore \quad \underline{E = 1429\ \text{V}}$$

It remains to find ω at this value of E with I_a equal to 230 A. For this we need a plot of E/ω against I_a. This can be plotted from the test results using

$$E = 1452 - 2 - I_a \times 0.3$$

I_a(A)	100	150	200	250	300
E(V)	1420	1405	1390	1375	1360
E/ω	390	540	538	717	756

From Fig. 7.21(b), $E/\omega = 690$, when $I_a = 230$

$$\therefore \quad \omega = E/(E/\omega) = 1429/690 = \underline{2.07\ \text{rad/sec}} \quad \text{or} \quad \underline{19.8\ \text{rev/min}}$$

When the armature circuit resistance is increased by 3 Ω then

$$E = 1500 - 230 \times 3.3 - 2 = 738\ \text{V}$$

and

$$\omega = 738/690 = \underline{1.07\ \text{rad/sec}} \quad \text{or} \quad \underline{10.1\ \text{rev/min}}$$

The inclusion of the 3 Ω series resistance has approximately halved the speed at which 150×10^3 Nm is developed.

Logically the following section would be on the series generator but this is such a rare device it is not worth including in a textbook of this type. The curious reader is referred to reference [5] for details of a series generator.

7.12 COMPOUND EXCITATION

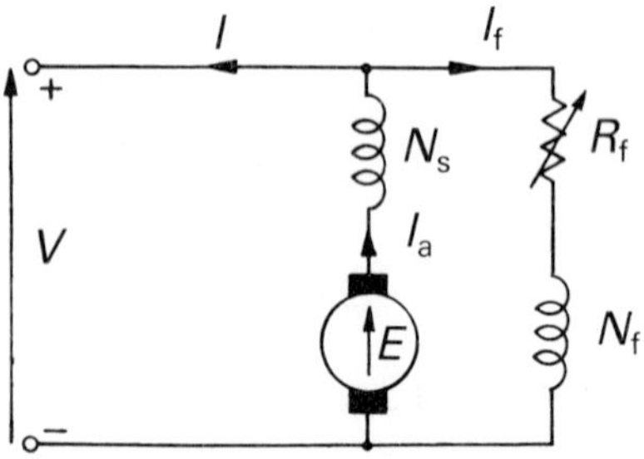

Fig. 7.22 Circuit diagram for a compound wound generator.

Figure 7.22 shows the circuit diagram for a compound wound generator. The poles carry two windings one of N_s turns and the other of N_f turns $N_f \gg N_s$ because $I_a \gg I_f$ and in addition the shunt-field winding has to dominate the field flux characteristic. The shunt-field circuit could also have been connected across the armature alone leaving the series field to carry the supply current rather than the total armature current as in Fig. 7.22. The series field can be connected in such a way as to aid the shunt field in producing flux or to oppose the shunt field. The former is referred to as 'cumulative compounding' and the latter as 'differential compounding'. They can be compared to positive or negative feedback respectively from the armature current to the flux per pole.

If we first of all consider the generator case the degree of cumulative compounding can be chosen so that at a particular load current the output voltage is the same as on open circuit. The series field m.m.f. has to offset the flux weakening effect of armature reaction and increase the flux per pole compared to open circuit in order to generate the greater E needed to offset the armature resistance volt drop. When the output voltage at rated current is the same as on no load then the machine is said to be flat compounded.

Differential compounding will always give lower output voltages at any load compared to the simple shunt generator. Fig. 7.23(a) shows the output characteristics of compound wound generators.

When applied to a motor cumulative compounding increases Φ for a given I_a and to produce a given T, I_a will subsequently be less than in the shunt motor ($T = K\Phi I_a$). The reduced I_a will result in a higher E for a given supply voltage. The machine will run more slowly than the shunt machine if the percentage increase in Φ is higher than the percentage increase in E. Bearing in mind that armature reaction weakens the flux per pole cumulative compounding can be used to offset any destabilising effects of armature reaction in applications subject to current overloads. Differential compounding in motors weakens the flux per pole and, like armature reaction effects, may lead to instability at high currents. Fig. 7.23(b) shows a set of speed–torque curves for compound wound motors.

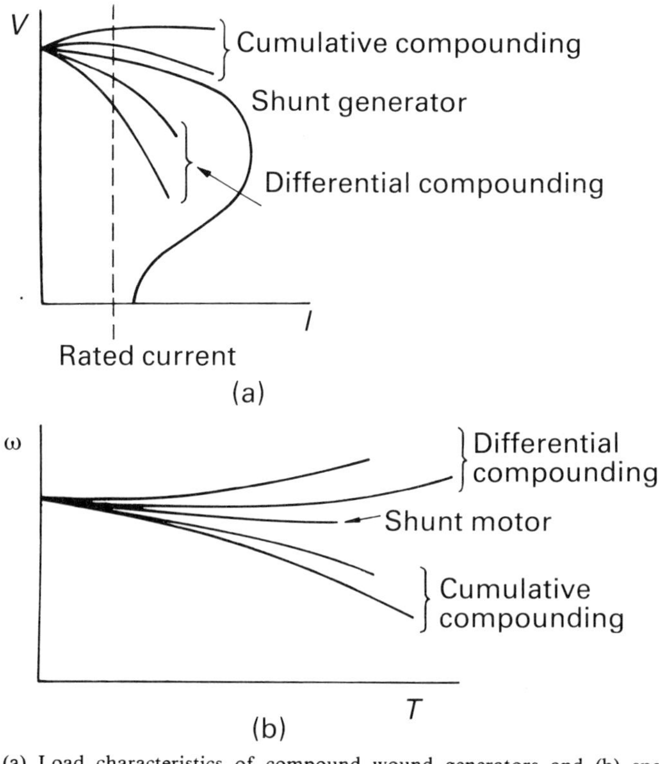

Fig. 7.23 (a) Load characteristics of compound wound generators and (b) speed/torque characteristics of compound wound motors.

It is of importance to note that a cumulatively compounded generator becomes a differentially compounded motor unless the series field winding connections are reversed when the armature current direction reverses.

Depending on the application this may turn a stable generator into an unstable motor and care should be exercised when using a compound wound d.c. generator in a situation where it may be called upon to motor, e.g. feeding power into a system where the supply voltage may increase above the machine e.m.f.

Problems on compound machines can be found in reference [5]. A general idea of the effects as explained above will suffice for this text.

7.13 THE UNIVERSAL MOTOR

Universal is a somewhat strong term to cover the fact that this motor can be used on both d.c. or a.c. supplies. It is essentially a d.c. series connected motor but it will have both the rotor and stator made from laminations. A motor designed for d.c. use only would not need a laminated stator. Provided the field flux and armature current reverse at the same time the torque will remain undirectional. Fig. 7.24(a) and (b) show the current, flux and torque direction for the positive and negative half-cycles. Because the current and flux are sinusoidal time functions the torque will pulsate at twice the supply frequency.

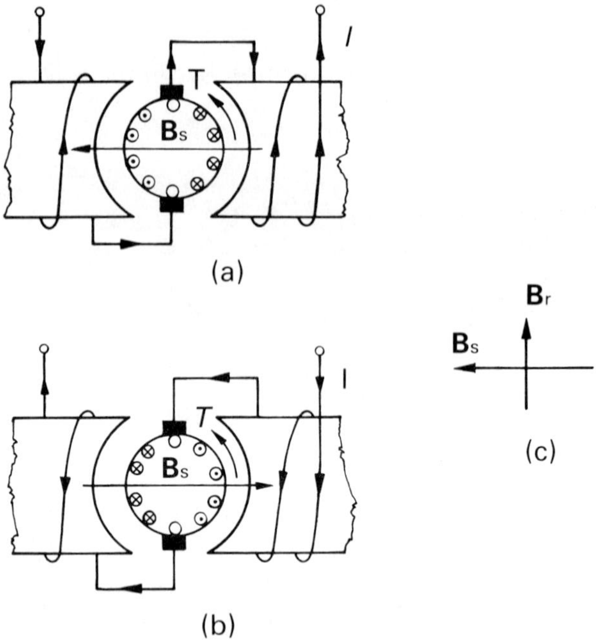

Fig. 7.24 (a) and (b) show unidirectional torque T as I and B_s reverse. (c) Flux density vectors for universal motors.

Fig. 7.24(c) gives an alternative way of looking at the torque in terms of the relationship between two stationary flux distributions B_s and B_r. B_r is set up by armature currents. If we consider only the fundamental space distribution components of B_s and B_r then

$$T = K'B_{s1}B_{r1} \sin\theta \qquad \text{(see equation (2.6))}$$

and since

$$\theta = 90^\circ$$

$$T = K'B_{s1}B_{r1}$$

Now if B_{s1} and B_{r1} are in phase then

$$T = K'\hat{B}_{s1} \sin \omega t \cdot \hat{B}_{r1} \sin \omega t$$

and this has an average value of $K'\hat{B}_{s1}\hat{B}_{r1}/2$ and a pulsating component at frequency 2ω.

The torque speed characteristics of the universal motor on a.c. are similar to the d.c. characteristics. The current and hence torque for a given applied voltage will be less due to the significant self-inductance of the field windings. This will have a similar effect to the series resistor effect in the d.c. case. The commutation problems are, however, more severe. The coils undergoing commutation do not 'cut' the changing field flux but they do 'link' it. There is therefore a transformer type e.m.f. induced in these shorted coils which is not related to the speed of rotation and this leads to increased sparking at the brushes. It is for this reason that a.c. series motors intended for use on mains frequency are small, typically < 1 kW. Where larger a.c. series motors are required the frequency must be reduced (Swiss National Railways use 50/3 Hz) or a more complex winding structure must be used to aid the commutation process.

7.14 STARTING D.C. MOTORS

When a motor is at rest there can be no back e.m.f. and if the supply voltage is connected to the armature there is only the armature resistance and brush drop to limit the armature current. The electrical time constant of the armature circuit is much less than the mechanical time constant of the rotor so that very high current levels can become established before the machine speeds up. In small machines, say < 1 kW, the armature resistance is disproportionately higher than in large machines and this coupled with their smaller mechanical time constants allows them to be started DOL (direct-on-line). This is also true for universal motors where the field inductance limits the maximum starting current. Other d.c. machines require some means of limiting the mechanical and electrical stresses during run up.

If a variable d.c. supply is available then the supply voltage can be reduced for starting purposes. Where a fixed voltage supply is available resistance must be inserted in series with the armature circuit for starting purposes. These starting resistors carry the full armature current but only for short periods of time. They usually take the form of a studded resistor successively switched out of circuit by a moving arm as shown in Fig. 7.25. The arm is also in contact with the incoming supply and must be of sufficient width to be in contact with the next stud prior to breaking contact with the previous stud. The rate at which the sections are switched out depends on the rate at which the machine speeds up. In manual starters this

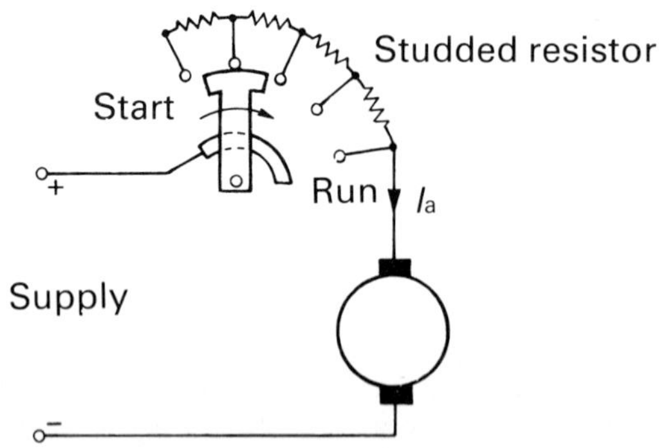

Fig. 7.25 Starting a shunt motor.

is under the control of the operator. Many machines are fitted with automatic starters in which the moving arm is replaced by a set of contactors which successively short-circuit the sections of the starting resistor under the control of a time switch or of some feedback system from the armature current or voltage. Fig. 7.25 does not show details of the field circuit. In the series machine it is of course in series with the armature and will carry armature current. In the shunt machine the field should be at its highest value to give maximum torque for any value of armature current. The shunt field rheostat should therefore be set at its minimum value and full voltage applied to the field before commencing the start-up procedure.

The series machine torque–speed curves show that at low load torques the machine will run at high speed. A d.c. series machine should not be run on no load since this will result in dangerously high speed values. To avoid this possibility many d.c. series motors have a centrifugal switch fitted to the shaft to disconnect the motor in the event of too high a speed. Care should be taken before starting a series motor to make sure that sufficient load torque is available to limit the motor speed. 'Starters' for d.c. motors incorporate many other features than the basic armature current limiters discussed above. Details of these features can be found in reference [5]. Power electronic 'packages' are also now available for the starting and control of d.c. motors.

7.15 WARD–LEONARD SYSTEM

Many industrial applications require precise control of torque and speed over a wide range including reversing, e.g. steel rolling mills. The Ward–Leonard system, named after its inventor and shown in Fig. 7.26, is a very suitable drive system for such applications. The prime mover may be an induction or synchronous motor or in a diesel-electric locomotive a diesel engine. The generator field can be reversed to reverse the motor and the combination of field and armature voltage control applied to the motor gives a very versatile system. In many applications the prime-mover and generator have been replaced by a controlled d.c. power source containing a three-phase rectifier and inverter. The inverter is necessary to allow for reverse power flow when the application calls for regenerative breaking, e.g. a crane or power shovel when lowering large loads. Where a quiet drive system is required, e.g. elevators in buildings, the drive motor and generator are retained since d.c. motors operating from controlled rectifiers tend to be noisy.

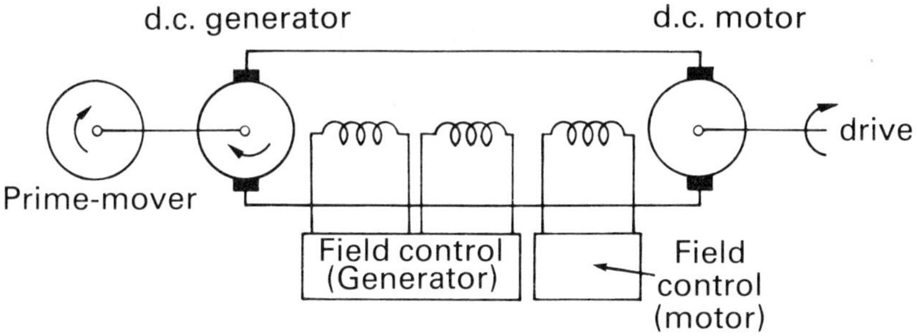

Fig. 7.26 The Ward–Leonard system.

7.16 APPLICATIONS

D.c. machines range in size from the miniature motors (run from dry cells) used in razors, cameras, etc., to several megawatt motors used in mining or process industries. Wherever a battery is the supply source, e.g. forklift trucks, delivery vans, the d.c. motor is the obvious drive motor. The starter motor in automobiles is a d.c. series motor and the windscreen wiper motors are permanent magnet d.c. motors. The great variety of speed–torque, current/voltage characteristics that can be achieved by altering the interconnection of field and armature circuits makes the d.c. machine a very versatile device. It has the great advantage of requiring only magnitude variations to achieve all control functions with no complications about frequency or phase as for the a.c. motor. Its great disadvantage is its higher cost and increased maintenance requirement to keep the commutator gear in good order.

D.c. generators are nearly always used in conjunction with a d.c. motor such as in the Ward–Leonard control system for elevators or rolling mills. Most variable voltage d.c. supplies now use controlled rectifiers rather than a d.c. generator but the output voltage is much less smooth.

D.c. series motors have a speed–torque characteristic well suited to traction applications and are found in most electrically driven trains. The universal motor is the standard choice for hand-drills, electric lawn-mowers, sewing machines, etc. where there is a need for a torque increase with speed decrease over the whole speed range.

7.17 CONCLUSIONS

Unlike the synchronous and induction machines we have not derived an equivalent circuit for a d.c. machine. We have none the less represented it as a variable voltage d.c. source in series with a resistor (R_a). The armature reaction effects on the flux per pole Φ are dealt with on a somewhat *ad hoc* basis in each problem due to the non-linearity of the effect. This contrasts with the synchronous reactance in the synchronous machine which represents the armature flux effects in the equivalent circuit while the magnetic path remains in a linear state.

The characteristics and application of the main types of d.c. machines have been covered in a fashion relevant to an elementary text of this nature. The reader who seeks or needs more detail is referred to reference 5.

PROBLEMS

1 The field winding of a d.c. machine is in two halves each of resistance 11.1 Ω and with the same number of turns. With the two windings connected in series and separately excited the open-circuit characteristic at 850 rev/min is as follows:

Field current (A)	1	1.5	2	3	4	5	6
Armature volts (V)	35	52	68	87	100.5	109	112.5

With the field windings connected as above and the machine connected as a shunt generator what will be the open-circuit e.m.f. at 850 rev/min? What is the critical field resistance at this speed? The two halves of the field winding are now connected in parallel and a field rheostat R_f inserted in series with this combination. The machine is run as a shunt connected generator at 425 rev/min. Calculate the value of R_f to give an open circuit voltage of 50 volts. Neglect brush drop and armature resistance.

Ans. 108 V, 35 Ω, 0.7 Ω.

2 During a test on a d.c. shunt motor a calibrated motor was used to drive it at a constant speed of 750 rev/min. Separate excitation was provided for the motor field and its armature was on open circuit. The test results were as follows:

Test machine				
Field current (A)	0.9	1.25	1.6	2.0
Armature volts (V)	180	210	228	245
Calibrated motor output (W)	660	741	850	1030

If the armature reaction is negligible what must be the resistance of the field circuit if the machine is to run as a shunt-connected motor at a speed of 750 rev/min drawing an armature current of 100 A from a 250 V supply? The armature circuit resistance is 0.14 Ω and the brush drop may be taken as 2 V. Estimate the efficiency of the motor under these conditions.

If the field resistance is set at 200 Ω at what speed must the motor run if the armature current remains at 100 A?

Ans. 144 Ω, 87.5%, 835 rev/min.

3 The speed of a d.c. motor driving a centrifugal pump is controlled by varying the armature voltage with the field current remaining constant. At full load the armature voltage is 500 V and armature current 136 A. If the load torque varies as the square of the speed calculate the voltage required across the armature in order to reduce the speed to 70.7 per cent of its full load value. The armature resistance is 0.25 Ω and lost-torque, armature reaction and brush drop may be neglected.

Ans. 347 V

4 The open-circuit characteristic for a shunt generator at rated speed is as follows:

Field current (A)	1	2.5	4.5,	7,	9	12	15	18
Armature volts (V)	20	210	340,	435,	490	550	584	610

The generator is then shunt-connected and run at rated speed. The field and armature circuit resistances are 44 Ω and 0.1 Ω respectively. Estimate the output voltage and current when the armature current is 400 A and (1) armature reaction is neglected, (2) armature reaction requires 1 A in the field to counteract its effect on the flux per pole.

Ans. 500 V, 388.6 A, 451 V, 390 A

5 A series motor runs at 580, 430 and 360 rev/min when drawing 20, 30 and 40 A respectively from a 200 V supply. The armature circuit resistance is 1 Ω. Find the speed at which it will run when drawing 35 A through a series resistance of 2.5 Ω from 200 V mains. Neglect brush drop.

Ans. 184 rev/min.

6 A railway car is being driven up an incline at 40 km/h by two identical d.c. series motors connected in parallel across a 500 V supply.

Tests on one motor at 40 km/h gave the following results:

Current (A)	100	130	160	190
Voltage (V)	460	551	607	638

The armature and field resistance together was 0.2 Ω.

If the resistance to motion of the car is assumed constant at what steady speed will the car continue to climb when one motor is disconnected from the supply?
(Hint: Plot a graph of gross torque (in arbitrary units) against current for one motor and assume lost torque only a function of current.) Neglect brush drop.

Ans. 31.3 km/h.

8

Other Machines including Magnetic and Linear Machines

8.1 INTRODUCTION

There is a host of special purpose machines not included in previous chapters. We shall have a superficial look at some of the more common machine types in this category without entering into any great technical detail of design and performance. The material in this chapter is therefore of a descriptive nature with no worked examples or problems.

8.2 RELUCTANCE MOTORS

In Section 2.7 we saw how a non-uniform air gap could give rise to a torque in a machine in which the rotor was neither a permanent magnet nor carried any current carrying conductors. Such a torque was called a reluctance torque since it arose from the variation of reluctance of the flux path with rotor position. Fig. 2.11(a) and (b) showed a very simple form of reluctance motor with a single coil on the stator carrying a current I_s. The basic torque equation is

$$T = \frac{1}{2} I_s^2 \frac{dL_s}{d\alpha} \tag{8.1}$$

where L_s is the self-inductance of the stator coil and α the rotor position with respect to a stationary reference axis. When I_s is a steady current this leads to the rotor taking up a position giving the minimum reluctance to the flux path. For a given current the torque is a maximum when $dL_s/d\alpha$ is a maximum and a great deal of ingenuity has gone into attemps to maximise this parameter in various rotor designs.

In order to produce rotation the machine needs to develop an average torque as α goes through one revolution. If we consider the variation of L_s with α for the machine shown in Fig. 8.1(a) and abstract the average and fundamental component of L_s then we can write

$$L_s = L_0 + L_1 \cos 2\alpha$$

This distribution is shown in Fig. 8.1(b). Now

$$\frac{dL_s}{d\alpha} = -2L_1 \sin 2\alpha \tag{8.2}$$

and if this is substituted into equation (8.1) we get

$$T = -I_s^2 L_1 \sin 2\alpha \tag{8.3}$$

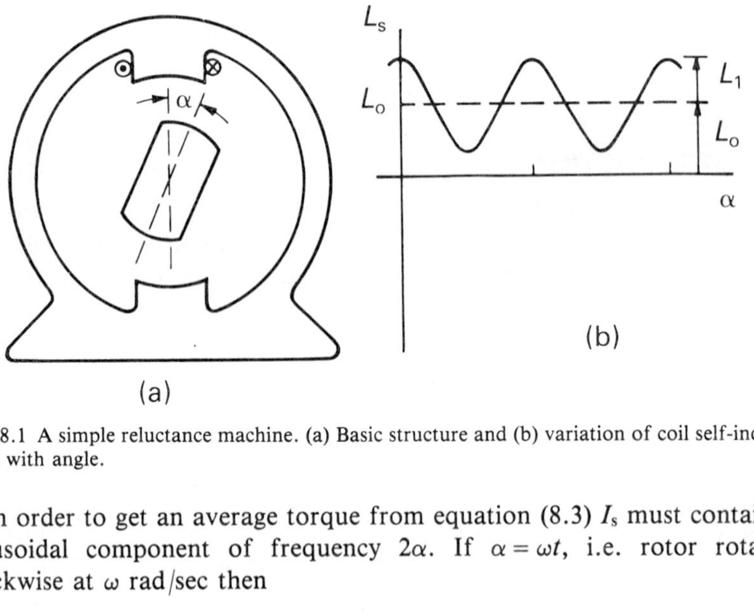

Fig. 8.1 A simple reluctance machine. (a) Basic structure and (b) variation of coil self-inductance with angle.

In order to get an average torque from equation (8.3) I_s must contain a sinusoidal component of frequency 2α. If $\alpha = \omega t$, i.e. rotor rotated clockwise at ω rad/sec then

$$T = -I_s^2 L_1 \sin 2\omega t \tag{8.4}$$

and $I_s = \hat{I}_s \sin(\omega t + \beta)$ will give an average torque in the direction of rotation for appropriate values of β. The direction of the torque will always be towards the nearest vertical position so that β must be chosen such that the torque as the rotor approaches the vertical is greater than the torque as the rotor moves away from the vertical. $\beta = 135$ or -45 can be seen to give maximum clockwise torque on this basis. This can be checked by substitution of the current expression into equation (8.4).

The motor shown in Fig. 8.1(a) is a single-phase reluctance motor and will have a pulsating torque component in addition to an average torque.

A constant torque can be derived from a machine with a three-phase distributed winding on the stator and salient 'poles' on the rotor. The rotor as before carries no winding. Salient pole synchronous motors which do carry a field winding develop a reluctance torque in addition to a torque arising from the interaction of rotor flux and stator currents. This increases the pull-out torque in comparison to a round rotor (non-salient pole) machine of the same size. When the stator does not have salient poles as in the case of a distributed three-phase winding the 'minimum reluctance' position is not a useful concept. It is still true that the rotor will try to line up with the 'poles' in the stator flux distribution, e.g. if we freeze the three-phase currents when i_a is a positive maximum ($i_b = i_c = \frac{1}{2}$ negative maximum) then the rotor will line up along the flux axis of coil A. The rotor will therefore move towards a position which maximises the flux through the rotor itself.

Instead of the stator windings being connected to a three-phase supply power electronics may be used to switch a d.c. supply from one phase to the next at an appropriate rate. Such a motor is called a switched reluctance motor and these motors may prove suitable for battery driven vehicles in preference to the present d.c. motors in use for this application. The simple robust structure of the rotor of a reluctance motor contrasts with the complex armature of a d.c. machine. The switching problems are taken care of in the stationary power electronic circuits.

The reluctance motor is a type of synchronous machine. It can only develop an average torque if there is a common frequency component in I_s^2 and $dL_s/d\alpha$. In order to start such a machine either a variable frequency supply is required or in a three-phase motor a closed rotor pole-face winding can be used to run the machine up as an induction motor to a speed at which the rotor pulls in to synchronism.

8.3 STEPPING-MOTORS [11]

When I_s in the machine of Fig. 8.1(a) was a d.c. current the rotor aligned itself in the minimum reluctance position. If the stator has more than one set of poles then a salient pole rotor can be made to align with any particular set of poles by passing a current through the windings on these poles. There is no need for the rotor to have the same number of poles as the stator as shown in Fig. 8.2(a).

When the coils on poles A and A′ are excited then the rotor assumes the position shown in Fig. 8.2(a). The rotor can be moved 30° in a clockwise direction by removing the excitation from poles A, A′ and exciting poles B, B′. If we wish to move counterclockwise, then excite poles C, C′ after A, A′. Hence by using a d.c. power supply and suitable switching circuits the rotor can be rotated in discrete 30° steps in either direction. If used in

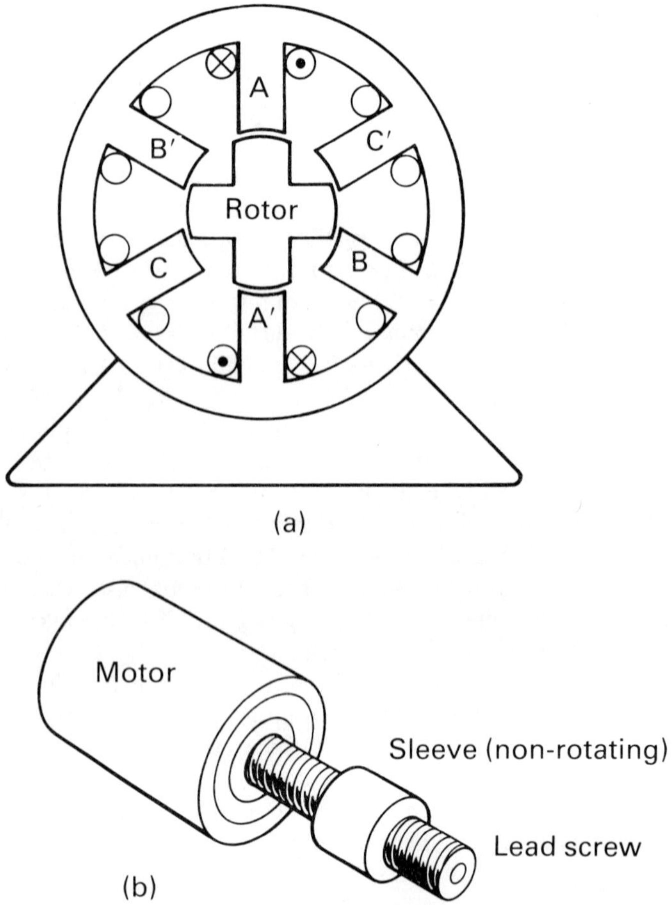

Fig. 8.2 The stepper motor. (a) The stator and rotor pole arrangements for a 30° step, (b) lateral positioning of a sleeve by use of a stepper motor and lead screw.

conjunction with a lead screw the number of steps through which the motor rotates can be used to accurately position a threaded sleeve riding on the screw as in Fig. 8.2(b). The error involved in positioning the rotor and hence the lead screw is independent of the number of steps. It will depend on the torque resisting the motion which will require that the rotor remains slightly out of line with the minimum reluctance position, where $dL/d\alpha$ is zero, in order to develop the necessary reluctance torque. The position error for moving round 1 step might be, say, 1° in 30°. The error if moving through 100 steps will still be 1° because the error only depends on the positional error in the last step. The errors are non-cumulative. Because of the accuracy attained, stepper motors may be used in an open loop positional control system, the output position being controlled only by the number of input pulses. The step length in the motor of Fig. 8.2(a) is quite large and other types of stepper motor exist, some incorporating permanent magnets in the rotor, which can give precision step lengths down to 0.9° where required.

The stepper motor requires a switching circuit which on receipt of an input pulse switches the d.c. supply to the 'next' phase. The 'next' phase will depend on the desired direction of rotation, e.g. in the machine illustrated in Fig. 8.2(a) the sequence ABC gives clockwise rotation and ACB gives counter-clockwise rotation. A logic controller, e.g. a microprocessor, needs to set the sequence and then the number of pulses required to give the desired rotation in the direction decided by the sequence instruction. The input pulses and logic signal for the sequence control are at low power levels, e.g. TTL circuits provide 5 V at 18 mA, while the switching circuit outputs are at much higher power levels suitable for the excitation levels in the phase winding of the motor, e.g. for a torque of 1.2 Nm a typical winding excitation is 3 A at 5 V.

Stepper motors of microwatt level are used in wrist-watches whilst in machine tools stepper motors with ratings of several tens of kilowatts may be found. They are widely incorporated in disc-drive systems for computers where the 'digital' nature of their input signals is ideal for interfacing with the digital electronics.

Reference [11] gives a much more detailed coverage of stepper motors should the reader wish to pursue the subject further.

8.4 DISC MOTORS

There are a number of applications in which axial length is restricted or in which there is a need for a low inertia rotor, e.g. sewing machines. Motors with disc rotors have been developed to satisfy these requirements.

Fig. 8.3 shows a typical structure for a permanent magnet disc d.c. motor. The armature is wound on an insulated disc with conductors on both sides (full and dotted lines) with the commutator at the centre. The need to make suitable connections to the commutator and interconnections between appropriate coil sides limits the inner diameter of the coil ends. The rotor disc is mounted between the permanent magnet poles and a soft iron backing plate which provides a return path for the pole flux. The rotor inertia is low since it does not contain iron but on the other hand there are two rather than one air gaps. Where there is no requirement for low inertia the backing plate can become part of the rotor but the plate will require to be laminated to avoid iron losses. Depending on the type of permanent magnet used the backing plate might be made of a ferrite material which will

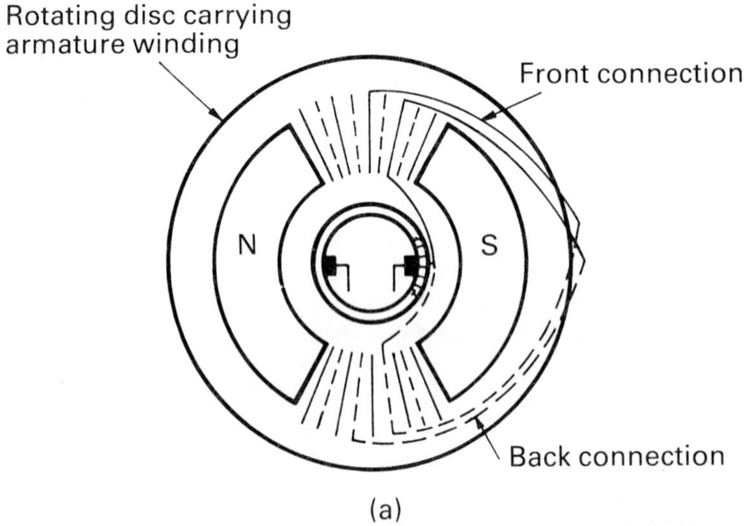

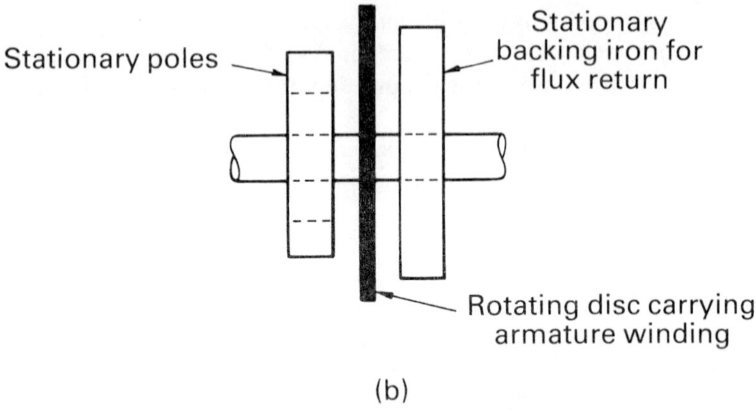

Fig. 8.3 An axial flux disc motor. (a) Axial view from the commutator end and (b) radial view.

decrease the rotor weight and minimise iron losses in the rotor. The disc armature can also be of the printed circuit type making for a very light, compact rotor.

D.c. disc type motors have been developed for use in car radiator fans and for direct mounting in the wheel hubs of electrically driven vehicles.

Fig. 8.4 shows a diagrammatic representation of a three-phase disc motor of the induction motor type. The copper disc, and backing iron plate, performs the same function as the squirrel cage in the radial flux motor of Chapter 6. Motors of this type have been used as pumps or stirrers with the disc mounted inside the fluid and separated from the stator by an imper-

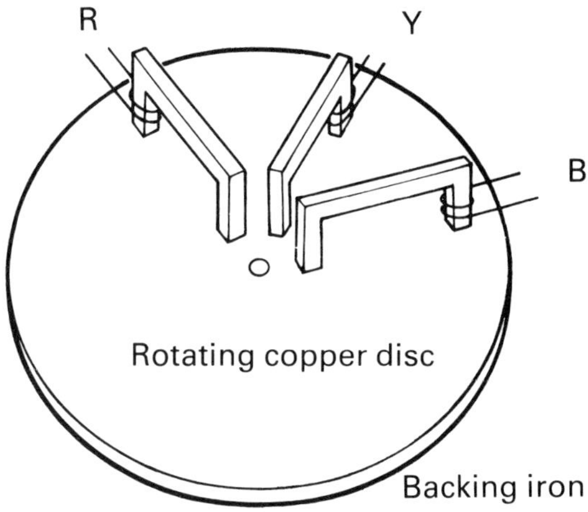

Fig. 8.4 Disc motor of induction type.

vious membrane. When a backing plate is mounted on the rotor disc there will be an axial force of attraction which the disc bearings must sustain. When used in fan or pump applications the thrust can be used to partially offset this force.

Perhaps the most common disc type 'motor' is the kilowatt/hour meter used in domestic premises to measure electricity consumption. Output torque, however, is not the objective!

8.5 LINEAR MOTORS

In theory all of the radial flux machines have linear counterparts. Indeed, we used linear developments of rotor and stator surfaces to facilitate explanations of machine operation. There is, however, a very important difference between the linear development of a radial machine and an actual linear machine. In the former the stator and rotor developments are the same length (apart from the approximation involving air-gap length) and after one revolution the rotor and stator are back in the same position with respect to each other. At a constant speed steady state conditions exist in such a machine. In the linear motor one member will be shorter than the other and at a steady speed the shorter member will be continually passing over a new part of the other member. At a steady speed in such machines the performance is greatly affected by the transient conditions which exist at the entry and trailing edges of the shorter member. This said, we shall not enter into a detailed analysis requiring the above refinements but it is none the less important to realise that a linear machine is not just an extension of a folded out radial machine. Reference [2] covers the necessary detailed analysis for the interested reader.

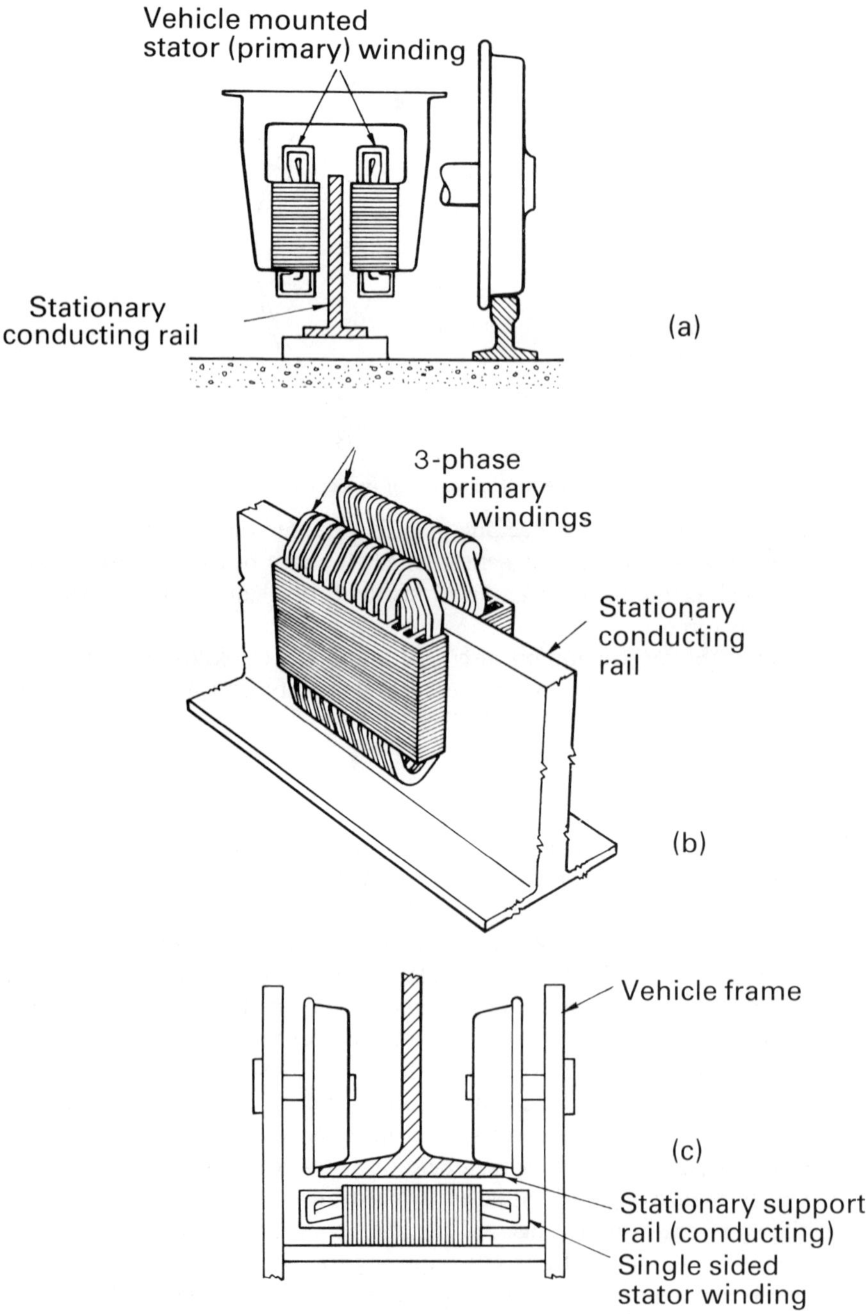

Fig. 8.5 Linear induction motors. (a) Double-sided 'short' primary winding, (b) detailed perspective view of primary windings and (c) single-sided 'short' primary winding.

The most common type of linear motor is the induction motor. In its usual form it consists of a 'short' stator winding which may be double sided as shown in Fig. 8.5(a) and (b) or single sided as shown in Fig. 8.5(c). The stator or primary winding is a distributed three-phase winding possibly fed from a variable frequency three-phase supply where speed control is necessary and the secondary (equivalent to rotor) winding is a conducting or 'reaction' rail which may be part of the structure in which the device is being used as in Fig. 8.5(c). The travelling flux wave set up by primary currents causes secondary currents as in a squirrel cage machine which interact with the primary flux to give a force on the primary in the *opposite* direction to the travelling wave.

The torque–speed curve is similar to that for the radial-flux squirrel-cage machine but the larger air gaps found in linear machines reduce the flux levels in comparison to radial machines. In the double-sided stator machine there is no lateral force between primary and secondary provided the secondary is placed centrally in the air gap. In the single-sided stator machine there is a lateral force of repulsion between the secondary and primary currents which may or may not be offset by an attractive force if the secondary is a ferromagnetic material.

This force of repulsion can be used to levitate as well as propel the secondary in a machine which has a short secondary as shown in Fig. 8.6. This type of machine has been referred to as a 'magnetic river' since the plate is swept along by the travelling flux wave set up by three-phase currents in coils ABCA—. The primary flux loops are completed through the air above the U section with the flux lines running at right angles to the direction of motion of the secondary. Such machines are called transverse-flux machines.

In addition to being used in cranes and trains as shown in Fig. 8.5 linear induction motors have been used in accelerators for rigs for testing vehicle performance under impact conditions.

The Japanese have used a linear synchronous motor to propel their high speed superconducting levitated trains. The superconducting coils on the train provide the field flux which interacts with the travelling flux wave produced by a continuous three-phase winding on the track. The vehicle speed is altered by changing the frequency of the currents supplied to the track windings. Although the track winding is expensive it overcomes the problem of transferring large amounts of electrical power to a vehicle moving at high speed (500 km/h). [6]

8.6 SUPERCONDUCTING MACHINES

When Onnes discovered the phenomenon of superconductivity in 1911 it was thought that superconducting coils immersed in liquid helium would be able to produce very large magnetic fields without the need to supply any power once the field was established. Subsequent workers soon discovered that the early type I superconductors lost their superconducting properties in relatively low magnetic field densities (1 Tesla) and it was not until the discovery of type II superconductors (niobium–titanium, niobium–tin) and their subsequent use in multifilament form embedded in a copper or bronze matrix that superconductors began to realise the promise of being able to produce high magnetic fields from self-sustaining current loops. Even now superconducting coils can lose their superconducting properties if subjected to relatively low level a.c. fluxes. Present day superconductors cannot be used in applications where they would have to carry power frequency currents or undergo rapid changes of current, e.g. in a d.c. machine armature. Their use is restricted to applications where a d.c. field flux is required and where the coil producing this flux can be protected from the flux variation caused by changing currents in normal conductors elsewhere in the machine. There is the added complication of immersing the superconducting coil in liquid helium. Present day superconductors of any technological significance require to operate in temperatures less than 10° K. Considerable advances have been made in recent years in the operation of liquid helium systems and this is no longer the major problem in superconducting machines.

The first application to come to mind would be the field in a d.c. machine. The cost of the refrigeration system could only be justified in a large machine and if the iron is retained then the flux density is limited to about 1.5 Tesla by the saturation properties of the iron path. The a.c. flux ripples

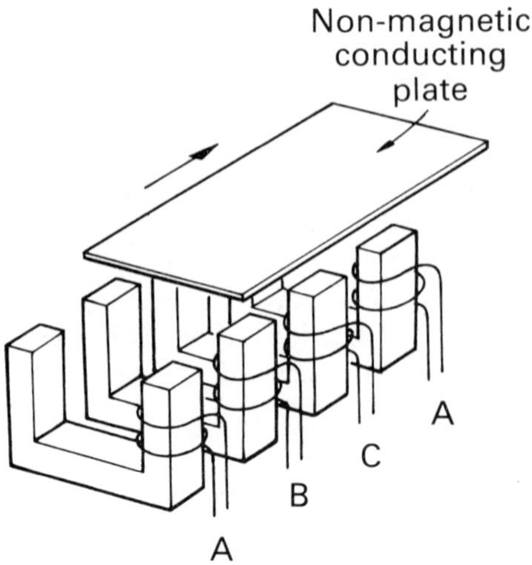

Fig. 8.6 Transverse flux linear induction machine ('magnetic river').

due to rotor teeth and commutation would probably destroy the field coil superconductivity and the coil would need to be shielded against flux changes caused by changing armature currents. Even if we removed the iron and used non-magnetic material to hold the field and armature coils the armature reaction flux changes would still destroy the field coil superconductivity. The conventionally designed radial configuration is clearly not suitable for a superconducting d.c. machine. Curiously enough the earliest type of d.c. machine, the Faraday disc, is the best configuration since the armature reaction flux loops produce no coupling with the main field coil. Armature current changes do not interfere with the field coil and furthermore the field coil does not experience any force from the armature currents. Since superconductors are extremely susceptible to slight changes in position this is another advantage of the Faraday disc machine. (At this point the reader may like to be reminded that the force interaction is

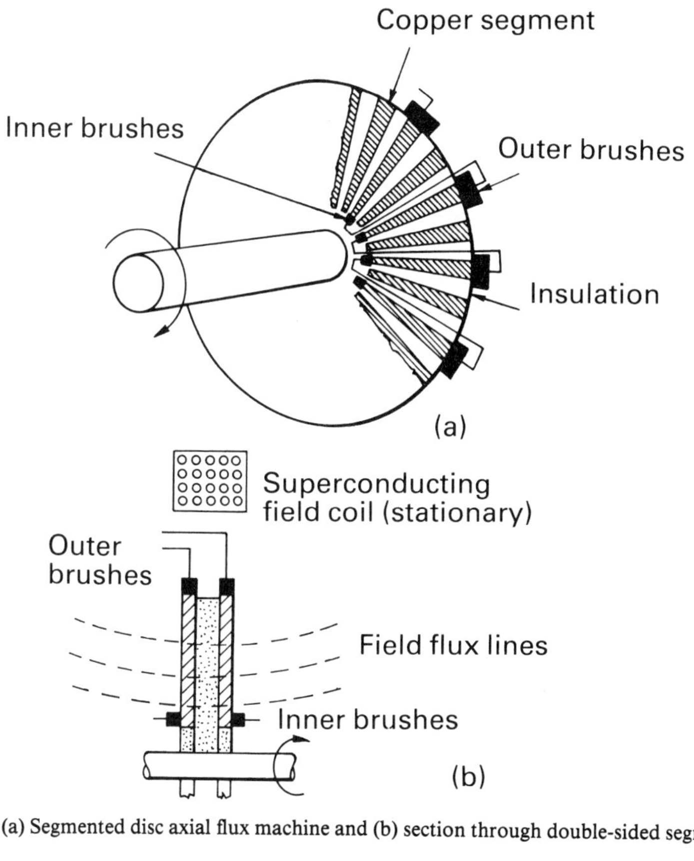

Fig. 8.7 (a) Segmented disc axial flux machine and (b) section through double-sided segmented disc.

between the fixed and moving parts of the armature current loop.) Since even in a fairly large disc travelling at high speed the e.m.f. along a radius is fairly small, segmented rotors with numerous rubbing contacts are used as shown in Fig. 8.7(a).

The fixed connections linking the brushes need to be well braced since they sustain the reaction torque. A double-sided disc may be used to make better use of the field flux as shown in Fig. 8.7(b). A 2.25 MW motor of this type has been successfully test run and there is considerable interest in the use of such motors in marine drives for large ships where the superconducting motor offers considerable space saving in comparison to a conventional diesel plus gearbox drive. The absence of iron in the machine, although it allows flux densities of the order of 3 Tesla to be achieved, means that the motor must be kept away from metalwork that would distort the field and sustain very significant forces.

A considerable research and development effort has gone into the design of very large alternators (about 2 GVA) using a rotating superconducting field coil. Such a system has the added complication of coupling a stationary helium refrigeration system to a rotating dewar housing the superconducting coil. The coil also has to be shielded against the low frequency (2 Hz) armature flux changes that occur during power swings in an interconnected power system. The rotor design is therefore exceedingly complex. The armature (stator) is also not without problems since the forces sustained by armature conductors carrying large current in a high field density are difficult to support in the absence of iron. There is also a need to shield the environment from the quite large alternating fluxes outside the machine. None the less the savings in weight for a superconducting field alternator in comparison to its conventional counterpart at this power level, in addition to the increased efficiency as a result of saving the field coil I^2r loss, makes the investigation worthwhile. (In a 600 MW conventional alternator about 3 MW are consumed in the field coil.) The need for very large alternators depends very much on the need for nuclear power stations. The latter would be ideally served by one or two large alternators of this type.

Large superconducting coils are being used by the Japanese in their high speed ground transport vehicle [6]. The same coil system is used to provide lift, guidance and supply the field flux for a linear synchronous motor which propels the vehicle over a track containing continuous three-phase windings. Lift is generated by the superconducting coil flux cutting a continuous array of closed conducting coils on either side of the track. The currents in these coils repel the field coil giving sufficient lift to keep a 33 ton vehicle 25 cm off the ground at 500 km/h.

8.7 CONCLUSIONS

This chapter has given a very brief description of other machines which do not fall into the conventional a.c./d.c. machine pattern. The reluctance, disc, stepper and linear motors are all in everyday use whereas the superconducting machines are still in the research/development phase.

9

Power Electronics

9.1 INTRODUCTION

The area of application known as 'Power Electronics' developed in the early 1960s as the thyristor or SCR (silicon controlled rectifier) as it was then known replaced the previously available controlled rectifiers such as thyratrons and mercury arc valves. The thyristor offered all the same advantages of a small robust semiconductor device which led to the transistor replacing thermionic valves in all but a few specialised areas. It had one disadvantage in comparison to the transistor in that once triggered into a conducting state it could not be turned off by a low power signal on the base or gate. It could only be turned off by reversing the voltage across the device long enough to force the forward current to zero. On the other hand its voltage withstand capability in the forward direction when not conducting and its peak inverse voltage capability were far superior to the transistor. It also had better thermal stability, freedom from secondary breakdown problems and an ability to carry fault currents long enough to blow protecting fuses. It truly was a power device which could be controlled electronically and designers were quick to incorporate it in new power switching circuits to control motors, induction heating furnaces and high voltage d.c. power transmission converter stations, etc. We shall deal only with the use of power electronics as applied to motor control. The treatment is mainly descriptive and only the main types of drive systems will be covered.

9.2 SEMICONDUCTOR SWITCHING DEVICES

Both thyristors and transistors are now used in power switching circuits and a brief account of their principal properties will be given. The physics of both devices are covered in most books on electronics and will not be dealt with here.

9.2.1 Diodes and Thyristors

The semiconductor diode has a *V*/*I* characteristic as shown in Fig. 9.1(b).

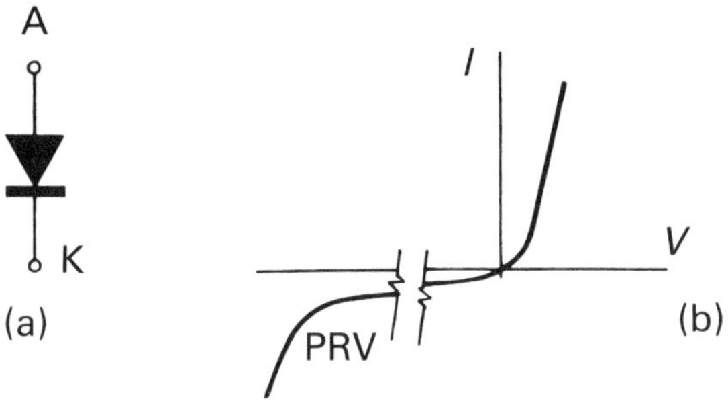

Fig. 9.1(a) Diode symbol and (b) typical *V*/*I* characteristic.

It is characterised by a low voltage drop in the forward direction of approximately 0.5 to 1 V over the permitted current range and very little current in the reverse direction until the peak reverse voltage (PRV) is reached at which point the diode breaks down. A power diode will carry several hundred amperes in the forward direction and have a PRV of several kV.

The diode may be used in simple rectifier circuits or it may be used in conjunction with thyristors in controlled rectifier circuits.

The thyristor symbol and *V*/*I* characteristic is similar to that for the diode with the exception of the gate electrode and its effect on the forward conduction. In the absence of any gate current the thyristor blocks the flow of current in both directions until either the PRV or forward breakover voltage is reached. Both of these voltages will be high in comparison to the 0.5 to 1.5 V volt-drop when the thyristor is conducting on the basic diode forward *V*/*I* curve. When the thyristor is forward 'biased', i.e. A is positive with

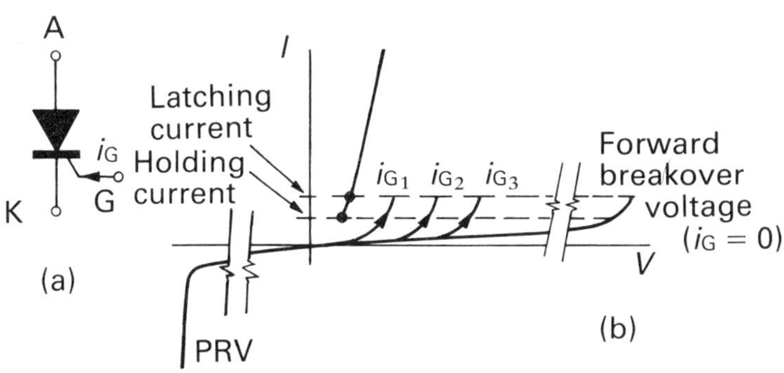

Fig. 9.2(a) Thyristor symbol and (b) thyristor *V*/*I* characteristic showing the effect of gate current.

respect to K, the thyristor can be 'triggered' or 'fired' into conduction at a particular forward voltage by controlling the gate current i_G in Fig. 9.2(a) and (b). The lower the A/K voltage the higher the gate current required to ensure conduction. The gate current is much smaller than the anode current, e.g. a 100 A r.m.s. thyristor would require a gate current in the range 50–200 mA to start conduction. Once the anode current has reached the 'latching' level the gate current can be removed and the thyristor remains conducting in an 'ON' state. When in the ON state the gate current cannot exercise any further control over the anode current. The latching current for the above thyristor would be about 50 mA. The thyristor will remain ON until the anode current drops below the 'holding' current level, about 20 mA for this thyristor. At this current the device reverts to a high resistance blocking state and provided the anode–cathode voltage remains low or reverses for a period known as the 'turn off' time then the thyristor blocks current flow in both directions and is said to be 'OFF'. The 'turn-on' time during which the latching current level must be reached is 2 μs and the 'turn-off' time is 20 to 100 μs for the thyristor under consideration.

Turning ON a thyristor when it is forward biased is relatively simple requiring only a short gate current pulse. Some precautions must be taken to limit dI/dt as conduction starts to avoid hot spots around the anode electrode junction. 500 A/μs is the limit for our example thyristor. A few microhenries in series with the anode is usually adequate. Care must also be taken not to apply too high a voltage or current to the gate electrode and this is usually achieved by using a Zener diode in parallel with the gate circuit and some resistance in series with the gate.

The thyristor once ON can only be turned off by the anode current dropping below the holding level and the anode–cathode voltage remaining low or reverse biased during the turn-off time. In a.c. circuits the reversal of the supply voltage will force the anode current to zero and reverse bias the thyristor at the same time. This would be called 'natural' or 'line' commuta-

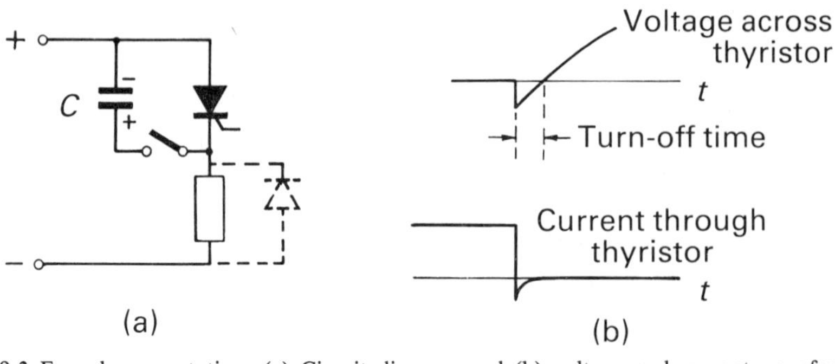

Fig. 9.3 Forced commutation. (a) Circuit diagram and (b) voltage and current waveform during and after commutation.

tion. When the supply voltage is d.c. as in Fig. 9.3(a) the anode current is forced to zero by connecting a charged capacitor across the thyristor such that the thyristor is reversed biased long enough to achieve turn off. This is called 'forced' commutation and in general involves much more complicated circuitry than the simple capacitor shown in Fig. 9.3(a).

The voltage and current waveforms during turn off are shown in Fig. 9.3(b) on the assumption that the load is resistive. If the load is inductive the voltage once the thyristor is OFF may increase rapidly due to $L(\mathrm{d}i/\mathrm{d}t)$ in the load. The capacitor C, depending on its size, will help to limit $\mathrm{d}V/\mathrm{d}t$ across the OFF thyristor but the inclusion of a 'freewheel' diode across the inductive load will greatly ease the problem. The inductor current can 'freewheel' through the dotted path in Fig. 9.3(a) once the supply path is broken by the OFF thyristor.

In addition to gate current turn-on a thyristor can be turned on by a high $\mathrm{d}V/\mathrm{d}t$ in the forward direction. Our example thyristor has a limit for $\mathrm{d}V/\mathrm{d}t$ of 500 V/μs. To avoid this type of turn on a 'snubber' circuit consisting of a parallel RC network is connected across the thyristor. $R = 10\,\Omega$ and $C \doteqdot 0.1\,\mu$F are typical values. Fig. 9.4 shows a typical thyristor circuit not

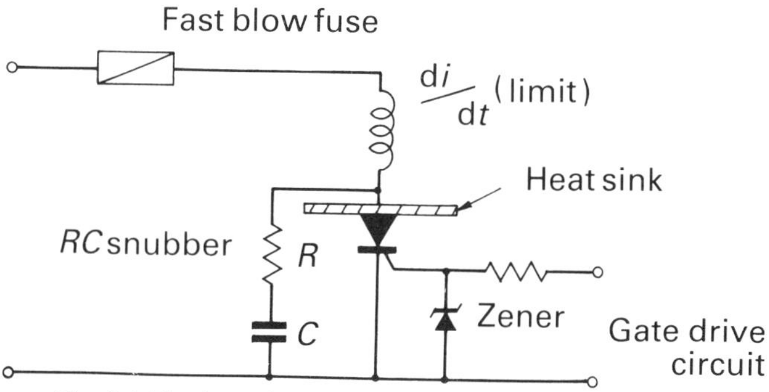

Fig. 9.4 Thyristor circuit showing main protective elements.

including any commutation circuit but containing all the necessary protecting elements. It contains two elements not mentioned so far. The heat sink is necessary because the ON volt drop of about 1 V coupled with the ON current generates a significant heat. The fast-blow fuse protects the thyristor from over-currents. The thyristor can withstand a significant current surge for a short period, 1500 A for 20 msec for our example thyristor, so that a fuse which blows in less than 20 msec at 1500 A will protect the thyristor. There will also be heat generated during switching when the VI product is quite large. When switching rates approach 150 to 200 times per second these switching losses may be larger than the ON losses mentioned previously.

In subsequent circuit diagrams the protection elements around the thyristor will not be included.

9.2.2 Power Transistors

Power transistors have a V/I characteristic of the type shown in Fig. 9.5(b). The current gain is of the order of 10 requiring a base current of 10 A for a collector current of 100 A. The volt drop across the device when conducting would normally be 0.5 to 1 V which is less than for the thyristor of corresponding rating.

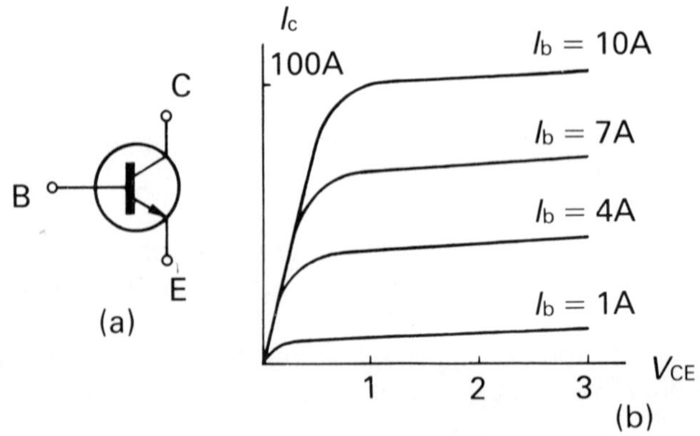

Fig. 9.5(a) An NPN transistor and (b) typical power transistor collector characteristic.

When not conducting the maximum V_{CE} in the forward direction would be of the order of 200 V and virtually zero in the reverse direction.

The transistor has the great advantage that it can be turned off by reducing the base current to zero and applying a small reverse bias to the base-emitter junction. On the other hand it requires a continuous base current when in the ON condition.

Fig. 9.6 shows a power transistor switching stage used in a PWM inverter. The base current drive for the two parallel 'output' transistors comes from

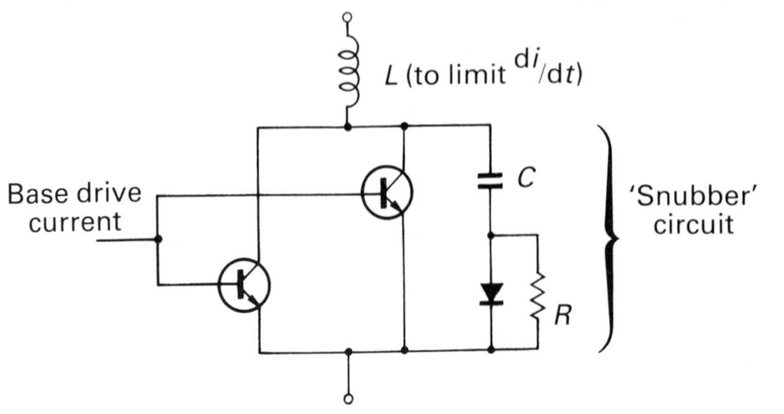

Fig. 9.6 A power transistor switching stage showing two parallel output transistors.

the emitter of a smaller power transistor. The series L and parallel snubber RC circuits are there for the same reasons as in the thyristor circuit, i.e. to prevent too rapid a voltage build-up of collector current at switch on and to prevent too rapid a voltage build-up across the transistor when switching off. The diode in parallel with R prevents the capacitor discharging rapidly through the transistors when they are switched on. R dictates the discharge rate when the transistors are ON. Note that there is no series fast-blow fuse. Transistors can only sustain over-currents for very short periods especially if the base drive current is insufficient to saturate the transistor at these currents. This will cause an increase in voltage across the transistor during the over-current leading to high heat generation in the device. If this persists for more than say 20 μs the device will be destroyed. No fuse can operate this quickly and electronic feedback protection circuits which detect dangerous operating conditions must rapidly shut off the base drive current. The transistor must also be mounted on adequate heat sinks which incorporate heat sensitive switches to prevent the base drive coming on if the heat sink temperature rises to a dangerous level. Transistors suffer from a condition known as 'thermal runaway' whereby the leakage proportion of the collector current (I_{co}) increases with temperature. Increased collector current causes increased temperature and so on until the device is destroyed unless the base drive is disabled. Thermal runaway only sets in above a certain temperature so that the device can be safely operated below this temperature. The bipolar power transistor is not as robust a device as the thyristor but its lack of commutation circuitry and faster switching speeds coupled to careful design are leading to its increased use in preference to thyristors where its power ratings are adequate.

The low current gain problem can be overcome by the Darlington connection shown in Fig. 9.7. Monolithic Darlingtons incorporate both transistors into one can showing only the terminals marked B, C and E in Fig. 9.7. Current gains of 100 can be achieved in this way but this may be accompanied by an increase in the ON state voltage.

Power MOSFETS with their insulated gate structure require negligible in-

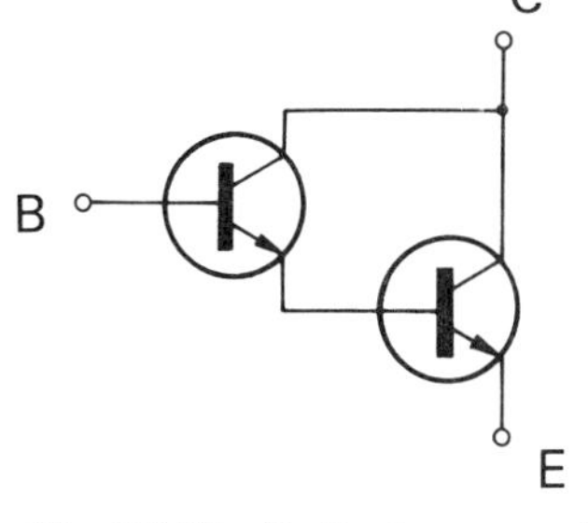

Fig. 9.7 The Darlington connection.

put gate current during the ON state. The gate drive circuit must be capable of charging the gate input capacitance if the inherent fast switching speeds of the device are to be realised. Even allowing for this requirement the gate drive circuitry is simpler than with bipolar transistors. The device also has a negative temperature coefficient for its ON current in that as it gets hotter its resistance increases tending to decrease current such that the *VI* product decreases. It is also less susceptible to secondary breakdown problems than the bipolar. Its one significant disadvantage is that the ON state volt-drop is three or four times that of an equivalent bipolar leading to significant ON state losses. On the other hand the switching losses are much lower and MOSFETS are therefore being used in applications where switching losses predominate, i.e. at 'high' frequencies, say, greater than 1 kHz. (The microwave engineer will wince at 1 kHz being called a high frequency!)

9.2.3 State of the art and future devices

Single thyristors are now (1982) available with a blocking voltage in both directions of 5 kV and a mean forward current of 100 A. Bipolar transistors, including Darlington, are available up to 1200 A at 120 V, up to 300 A at 400 V and up to 50 A at 700 V. Power MOSFETS are available up to 30 A at 450 V.

At high powers the thyristor is unrivalled, while at high frequencies, 5–30 kHz, the MOSFET is the obvious choice. Below 5 kHz the lower ON state resistance of the bipolar tells in its favour and as the frequency drops below about 200 Hz the thyristor comes into the reckoning again particularly if the power levels are high.

Power MOSFETS are mostly used in 'switching mode' power supplies to give smooth, regulated, d.c. power supplies. These are not required in machine drives but will be covered briefly in Section 9.4.

Bipolar transistors and thyristors are used in inverters, i.e. variable frequency, variable voltage a.c. power supplies. Transistors are rarely used above ratings of 100 kW but thyristor inverters are available up to 10 MW at 120 Hz. Thyristors are also used as controlled rectifiers in a.c. to d.c. converters.

Two thyristors mounted in parallel, but pointing in opposite directions, are available in the same can and may be fired by the same gate signal. Such a device is called a triac and is mainly used as an ON/OFF switch in a.c. circuits.

There are in addition several new semiconductor devices coming into contention such as the ASCR (asynchronous silicon controlled rectifier) the GTO (gate turn off thyristor) and FCT (field control thyristor). There is the more distant prospect also of a Ga As FET (Gallium Arsenide FET) which will have a much lower ON resistance than present power MOSFETS. For further information on these devices consult references [7] and [8].

9.3 A.C. TO D.C. CONVERTERS (RECTIFIERS)

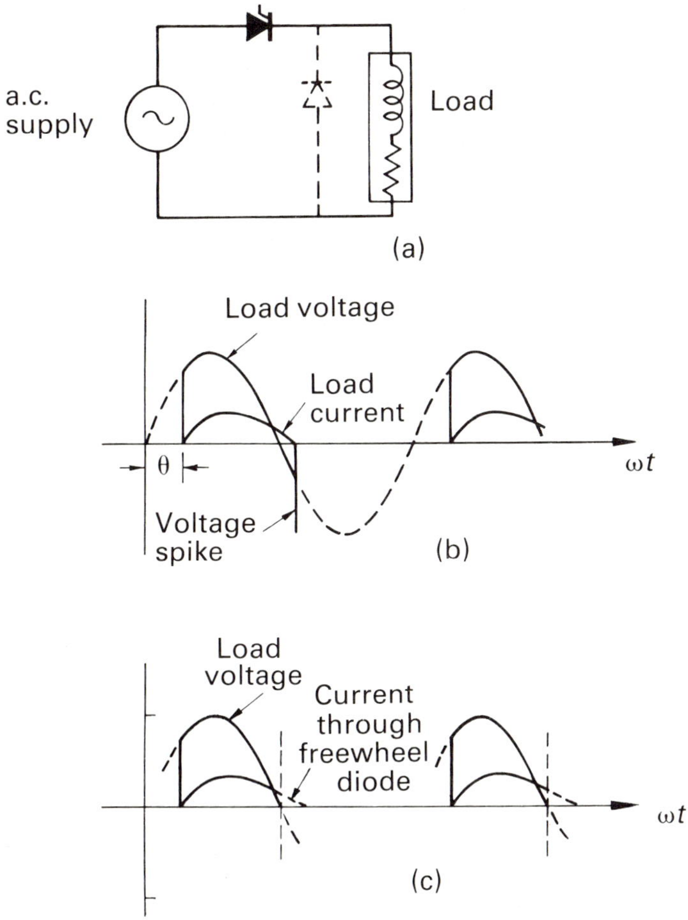

Fig. 9.8 Single-phase thyristor controller. (a) Circuit diagram, (b) load voltage and current waveforms, (c) load voltage and current waveforms with freewheel diode in circuit.

Fig. 9.8(a) shows a single-phase circuit in which the mean level of load voltage is controlled by firing the thyristor at some angle θ following the supply voltage zero. The current consists of an exponential component plus an a.c. component. The exponential component is normally short-lived in comparison to a half-cycle of power frequency. Once fired the thyristor remains in conduction until reverse biased, i.e. when the voltage across it reverses. This does not occur at the supply voltage reversal due to the presence of an inductance in the load. The simplest way to see when the thyristor voltage reverses is to look at the current waveform and observe where it goes through zero. As the current approaches zero and drops below the holding current level the current begins to fall off more rapidly than a sinusoid and this gives rise to an $L(\mathrm{d}i/\mathrm{d}t)$ term resulting in a voltage spike as shown in Fig. 9.8(b). The mean load voltage and current will be the average values of the waveforms in Fig. 9.8(b). Higher average values may be obtained for the same firing angle θ if a freewheel diode is included as shown dotted in Fig. 9.8(a). It serves two purposes. First, it 'line commutates' the thyristor immediately the supply voltage goes negative, i.e. in Fig. 9.8(a) there is a path for reverse current through the supply, freewheel diode and thyristor when the supply voltage reverses and this turns off the thyristor. Secondly, the freewheel diode takes up the load current when the thyristor switches off and allows the current to decay exponentially. Fig. 9.8(c) shows the load waveform in this case and it can be seen that the average values are higher.

A simple phase angle controller of this type will clearly draw harmonic currents from the supply and further, the fundamental current component always lags the supply voltage as a result of the delay angle θ. The power factor is always lagging.

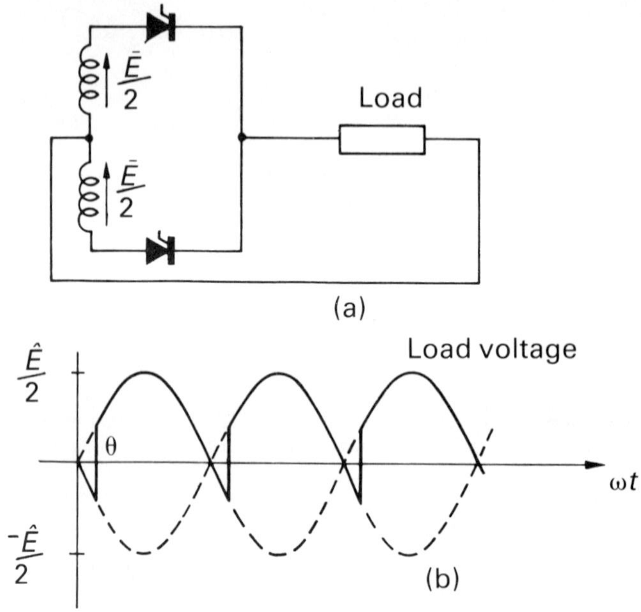

Fig. 9.9 Full-wave thyristor rectifier. (a) Circuit diagram and (b) load voltage waveform.

Fig. 9.9(a) shows a full-wave controlled rectifier with each thyristor being fired θ degrees after its supply voltage zero. Note that the firing of a thyristor automatically shuts off the other thyristor by reverse biasing it. The load voltage waveform therefore jumps from one voltage profile to the other when a thyristor is fired. If it was desired to prevent the load voltage going instantaneously negative then a freewheel diode could again be used across the load. The diode would commutate each thyristor as its supply voltage reversed and carry the load current (for inductive loads) until the next thyristor fired. The waveform in Fig. 9.9(b) assumes continuous conduction which implies an inductive load since conduction continues after the supply voltage goes through zero.

A three-phase half-wave rectifier is shown in Fig. 9.10(a) and the load voltage waveform in Fig. 9.10(b).

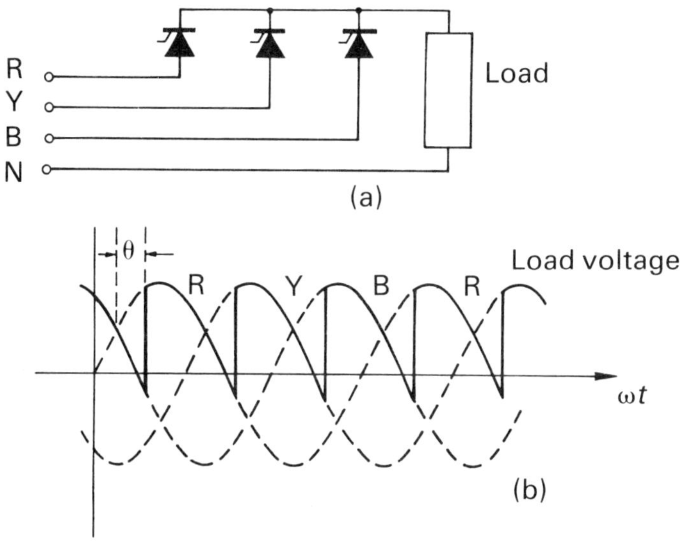

Fig. 9.10 Three-phase halfwave thyristor rectifier. (a) Circuit diagram and (b) load voltage waveform.

Note that the delay angle θ is no longer measured from a zero crossing but from a crossing of two of the voltage waveforms. Once again the firing of a thyristor shuts off the previously conducting thyristor by reverse biasing it.

For the three rectifier types shown the ripple voltage is least for the three-phase half-wave and could be improved still further by going to a three-phase full-wave or bridge connection. In all cases the fundamental current in a given phase lags the phase voltage giving rise to a lagging power factor.

In order to avoid a lagging power factor and supply harmonics some loads such as machines may be suitable for 'burst-firing' whereby the thyristors are fired with zero delay for several cycles and then not fired at all for several cycles.

Fig. 9.11 shows a symbolic representation of a controlled rectifier which

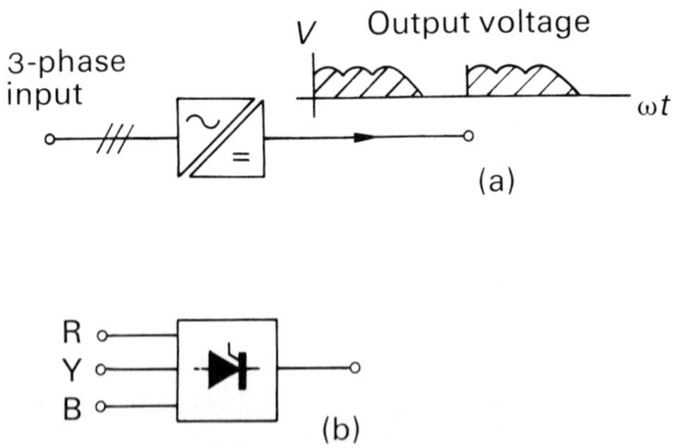

Fig. 9.11(a) Symbolic representation of a three-phase controlled rectifier used in a 'burst firing' mode and (b) Alternative symbol for a three-phase controlled rectifier.

converts incoming three-phase to d.c. by means of burst-firing. Fig. 9.11(b) is another widely used representation for a three-phase controlled rectifier.

9.4 D.C. TO D.C. CONVERTERS

There is a need to vary the d.c. voltage applied to a load device when the available power supply is at a fixed d.c. voltage. Battery driven vehicles is an obvious case in point where there may be a need to vary the armature voltage in order to derive the desired speed/torque relationship. Wasting power in a series resistor is not acceptable when energy is at a premium as in this case. The solution is to break up the supply voltage into periods of full voltage and periods of zero voltage. By varying the mark space ratio

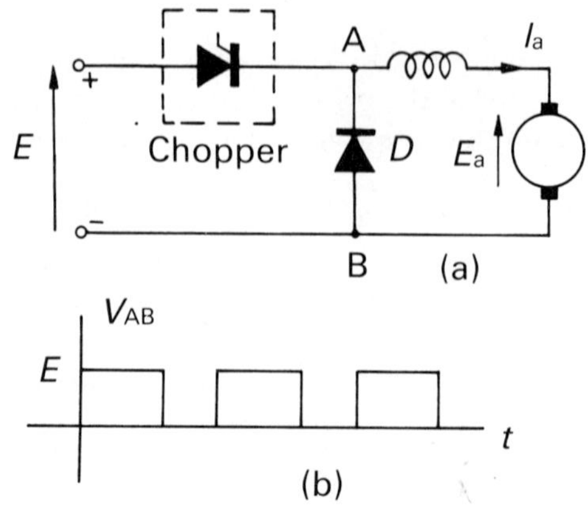

Fig. 9.12(a) A thyristor chopper driving a series motor and (b) chopper output voltage.

of these two periods the mean d.c. level can be set to the desired level. When thyristors or transistors are used in this mode of operation the device is called a 'chopper'. The chopper frequency is likely to be in the region of 250 Hz to 1000 Hz for a thyristor chopper but may be higher for bipolar and GTO devices. Fig. 9.12(a) shows the basic form of a chopper circuit feeding a d.c. series motor. No details are shown of the firing, commutation and protection circuits. The voltage applied to the machine, V_{AB}, is shown in Fig. 9.12(b). The current I_a continues to flow through the freewheel diode D when the chopper thyristor is OFF. The chopper circuit shown in Fig. 9.12(a) only operates for E_a and I_a in the direction shown. This would be called a single quadrant chopper. In order to allow for any combination of current and e.m.f. direction a four quadrant chopper with zone of operation as shown in Fig. 9.13(b) is required. The circuit for such a chopper is

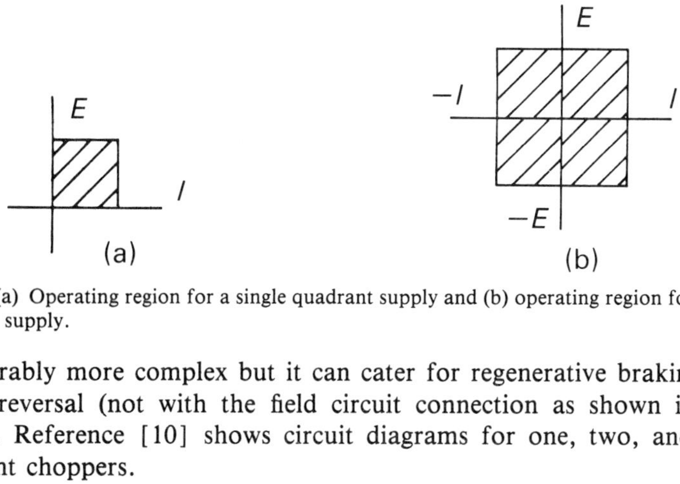

Fig. 9.13(a) Operating region for a single quadrant supply and (b) operating region for a four quadrant supply.

considerably more complex but it can cater for regenerative braking and motor reversal (not with the field circuit connection as shown in Fig. 9.12(a). Reference [10] shows circuit diagrams for one, two, and four quadrant choppers.

In low voltage choppers used in conjunction with batteries bipolar transistors would almost certainly be used since their ON state voltage is less than any of the other devices. Bipolars can also now be used with d.c. supply voltages up to 400 V but above this value either GTOs or ASCRs would be used. This level of voltage is common in d.c. traction applications where the supply rail will be around 400 V.

The switched mode power supply (SMPS) can be seen as a derivation of the chopper. The input voltage is normally full-wave rectified single-phase a.c. This produces an input voltage of around 300 V which will normally have been 'lightly smoothed'! This voltage is then chopped at around 20 kHz by a bipolar, MOSFET or GTO device, transformed down to a suitable level to allow an output d.c. voltage of say 0–60 V and then smoothed using LCR filters to give a nearly constant d.c. voltage. The high frequency involved allows the transformer and LCR smoothing elements to be much smaller than for a 50 Hz rectified power supply. The output voltage can be varied by adjusting the mark–space ratio of the switch and by suitable feedback to the switch control the output voltage can be kept within very close limits as both the output current and input d.c. voltage vary. The latter is due to the inadequate smoothing of the rectified 50 Hz. There is no call for this degree of precision on voltage levels to small machines but the SMPS is replacing the older series regulated d.c. power supplies.

9.5 D.C. TO A.C. CONVERTERS (INVERTERS)

This is the area which has seen the greatest growth with respect to machine drives. By being able to convert d.c. to variable frequency a.c. the prospect of electronically endowing the squirrel-cage induction motor with any desired torque–speed curve was extremely attractive. The prospect has not yet been realised but significant progress has been made.

Line commutated inverters have been in use on high voltage d.c. transmission systems for some considerable time. They rely on the existence of an already existing a.c. supply into which the d.c. power is to be fed via the inverter. The a.c. supply voltage can be used to commutate the thyristor and no additional commutation or snubber circuits are required. Such inverters are invariably three-phase devices but we shall look at the principle involved by examining a single-phase half-wave inverter.

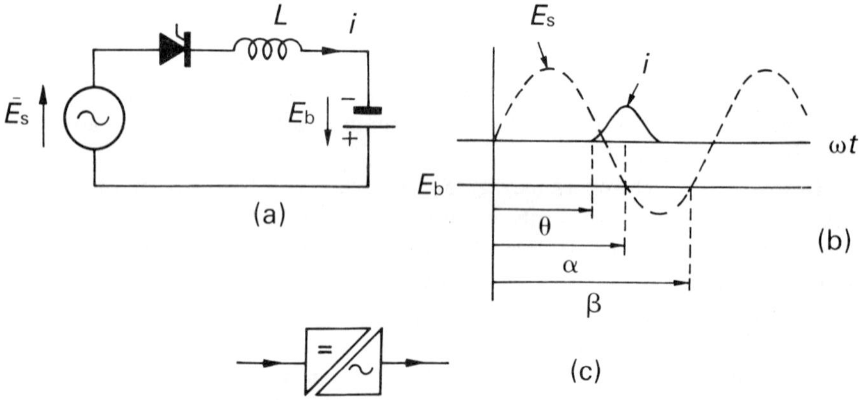

Fig. 9.14 A single-phase inverter. (a) Circuit diagram, (b) supply voltage and current waveform, (c) symbolic representation of inverter.

Fig. 9.14 shows the circuit with the load being replaced by a d.c. source of voltage E_b. The object of the exercise is to get i flowing while E_s is in its negative half-cycle. The thyristor should therefore be fired while it is forward biased but close to 180° on the supply voltage waveform. The bias voltage remember is not just E_s but $(\hat{E}_s \sin \omega t + E_b + L(\mathrm{d}i/\mathrm{d}t)$. After angle α on Fig. 9.14(b) $\hat{E} \sin \omega t$ is more negative than E_b and i will start to decrease. Conduction will continue until $\hat{E} \sin \omega t$ is more negative than $E_b + L(\mathrm{d}i/\mathrm{d}t)$ at which point conduction ceases until the thyristor is fired again in the following cycle. Conduction must cease before angle β in Fig. 9.14(b) because thereafter the thyristor is forward biased again. The average ei in the supply is negative due to e being negative over the greater part of the conduction period. This indicates energy input to the a.c. 'supply' from the d.c. voltage source.

Practical line commutated inverters use full-wave circuits, probably polyphase and it is possible to get a substantially constant current from the d.c. source with the current path being switched through a thyristor connected to a phase in which the supply e.m.f. is on average negative during the thyristor conduction period. The circuit and waveform diagram for a polyphase line commutated inverter are beyond the scope of this text. Greater detail can be found in Reference [10].

Fig. 9.14(c) shows a diagrammatic representation for an inverter. Fig. 9.11(b) is another possible diagrammatic representation since it is by control of the firing angle that the same device can convey power in either direction. Note that the current direction cannot change and that the power reversal is achieved by timing the current pulse to occur at an appropriate part of the a.c. supply voltage cycle.

Line commutated inverters can be used with synchronous machines where the machine e.m.f. can be used to commutate the inverters. The generated e.m.f. is not dependent on the presence of stator current and depends only on rotation and rotor field current. In the case of the induction motor there is no such e.m.f. present to commutate the inverter and forced commutation is necessary. Whereas the line commutated inverter will almost always use thyristors because of their more robust qualities, the forced commutation inverter is more likely to use transistors or GTOs unless the power levels dictate the use of large ASCRs.

Fig. 9.15(a) shows a typical arrangement for a transistor inverter in which the transistors operate as on–off switches with 120° between the switching periods in each phase. The line voltages consist of six steps and this type of inverter is a 'six-step' or 'quasi-square wave' inverter. The diodes in Fig. 9.15(a) provide a path for load current when the load current direction is not consistent with the forward direction of the ON transistor, e.g. if the phase R current lags the phase R voltage then when the top transistor on phase R switches off and the bottom transistor switches on the current flows through the bottom diode until the current changes sign whereupon it flows through the bottom transistor. The diodes would be necessary no matter what semiconductor device was used as the switch and are not just a feature of the protection circuitry for transistors. They do of course prevent an ON

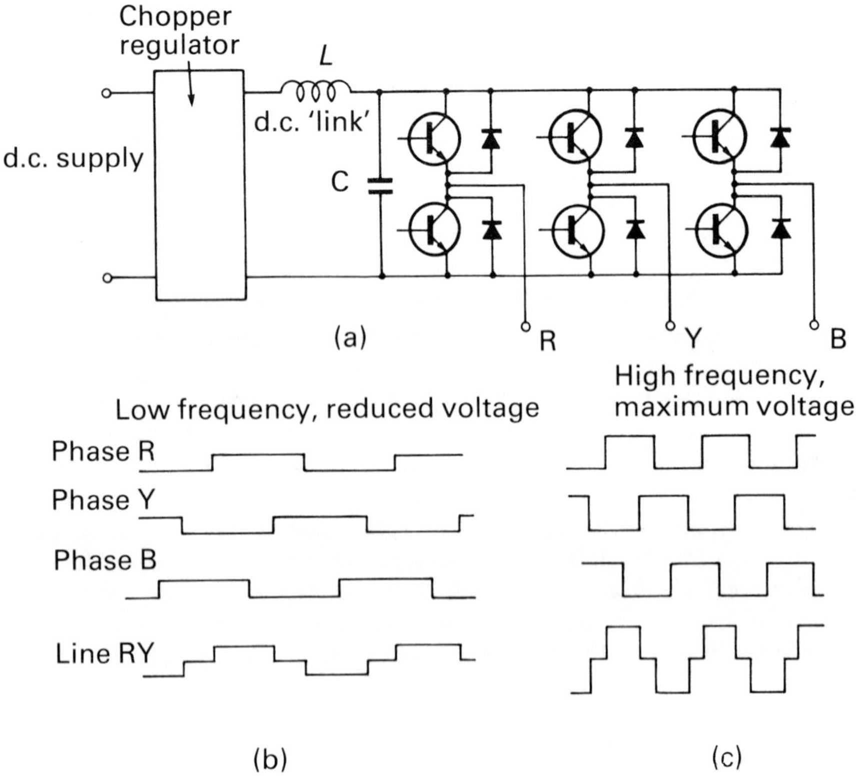

Fig. 9.15 A 'six-step' transistor inverter. (a) Circuit diagram, (b) output voltage waveforms at low frequency, (c) output voltage waveforms at high frequency.

transistor from being reverse biased! The d.c. supply rails to the inverter are referred to as the 'd.c. link' and most inverters require that the d.c. voltage is variable. The reason for variation of the d.c. voltage is that in order to keep the maximum flux density in a machine constant the voltage amplitude should be proportional to frequency. ($E \propto (\mathrm{d}\phi/\mathrm{d}t) \propto \omega_\mathrm{m} \cos \omega t$.) The chopper varies the d.c. voltage in accordance with the operating frequency as shown in Fig. 9.15(b) and (c).

This type of inverter suffers from two disadvantages which become more serious as power levels increase. The power transferred has to pass through two semiconductor systems, the chopper and the output switches, with consequent losses in both. There are also significant harmonic components in the voltage waveform, and at low frequency (about 10 Hz) both the electrical and mechanical time constants of the machine allow significant harmonic currents to flow which give rise to vibration and losses. On the other hand this type of inverter minimises the switching requirements.

When the switches are capable of faster operation then the two disadvan-

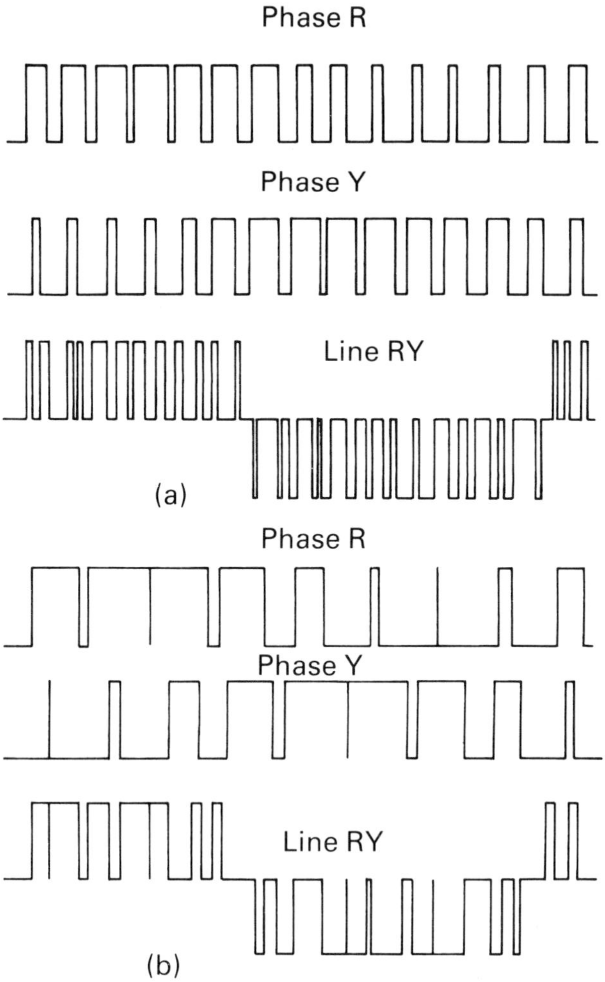

Fig. 9.16 PWM inverter output waveforms. (a) At low output frequency and reduced output voltage and (b) at high output frequency and high output voltage.

tages mentioned above can be avoided by using a pulse-width-modulated inverter (PWM). The output switches can take over the duty cycle of the chopper switches to produce waveforms of the type shown in Fig. 9.16(a) and (b). The 'chopping' frequency would typically be in the range 250 Hz to 1 kHz and the number of duty-cycles in one cycle of the operating frequency will be high at a low output frequency as in Fig. 9.16(a) becoming fewer as the output frequency increases. The mark–space ratio of any particular duty-cycle increases with frequency to give an amplitude increase with frequency. When the output frequency coincides with the chopping frequency the PWM inverter becomes a quasi-square wave inverter and the mark–space ratio is 1. However, this is at a frequency where the machine time constants will allow much less response to harmonics in the voltage waveform.

The real advantage of the PWM inverter is at low output frequencies where the next lowest frequency to the desired output frequency is shifted up to the sidebands around the chopping frequency caused by the output frequency, e.g. at an output frequency of 10 Hz and a chopping frequency of 500 Hz then the lowest parasitic frequency component is at (500 – 10) = 490 Hz. Even when the output frequency and chopping frequency are closer together microprocessors using look-up tables to control the output switches can eliminate any desired harmonic or minimise a number of harmonics. The PWM type inverter will therefore decrease harmonic losses in the machine but will increase switching losses in the inverter (including commutation losses if thyristors are used). The choice of chopping frequency will depend on a careful balance between these two losses.

9.6 THE CYCLOCONVERTER

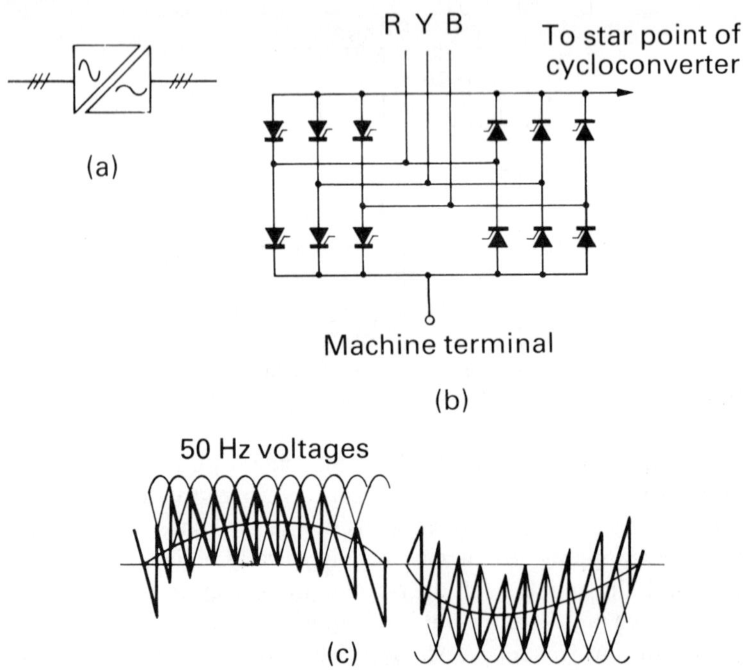

Fig. 9.17 The cycloconverter. (a) Symbolic representation, (b) one phase of a four-quadrant cycloconverter, (c) low frequency output built up from sections of 50 Hz waveforms.

The cycloconverter is a static (it does not rotate) frequency changer which can handle power flows in either direction. Fig. 9.17(a) shows the symbolic representation and Fig. 9.17(b) shows the circuit for one phase of a four-quadrant converter.

Each phase of the converter is fed from a separate three-phase transformer in this connection and the phase voltage is made up by picking off appropriate parts of successive 50 Hz voltages as in Fig. 9.17(c). The output frequency is limited to about 20 Hz for a 50 Hz input. The cycloconverter control and firing circuits are complex and usually incorporate feedback signals from the machine rotor position and stator current. On the other hand it is line commutated and the current waveform tends to be reasonably sinusoidal. It is used to control low speed synchronous motor drives of large power rating (up to 30 MW) and in induction motor speed control systems which utilise slip-power techniques.

9.7 VARIABLE SPEED D.C. DRIVES

Power electronic 'packages' are now available to suit the complete range of d.c. machine types and ratings. Feedback signals from the shaft speed and torque can be used to produce constant speed or constant torque drives or, with the aid of a microprocessor, any desired speed/torque curve. The package will include a controlled rectifier, or a simple rectifier followed by a chopper, to supply the armature circuit and will have independent control for the field circuit in the shunt motor case. The simple rectifier and chopper give a better input power factor on the a.c. side but will be more expensive.

Choppers are also being widely used in transit systems where the original machines were fed from a constant voltage d.c. rail with voltage control being achieved by inserting series resistance in the armature circuit. Chopper control is much more energy efficient. The chopping frequency should be as high as possible in order to minimise harmonic losses in the machine and the size of any filter components on the mains side. The frequency is limited by the commutation time in thyristor choppers and increased switching losses in both thyristor or transistor choppers. A typical chopping frequency is around 500 Hz but this could be much higher in a transistorised chopper. Choppers are rarely used for large machines say over 100 kW where a controlled rectifier is more likely. The 'static' Ward–Leonard system is a good example of an application to large machines. Fig. 9.18(a) shows a four-quadrant system which includes driving, regenerative braking and speed reversal by means of reversing the current in the armature circuit. The two converters in parallel must never be fired together otherwise a short-circuit occurs. Armature current feedback is required to the control electronics to sense the current zero at which

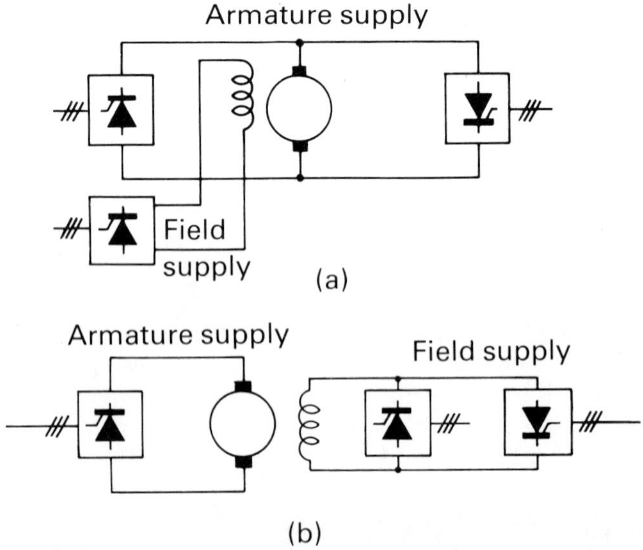

Fig. 9.18 A variable speed d.c. drive system. (a) A four-quadrant armature supply combined with a single quadrant field supply and (b) a two-quadrant armature supply combined with two two-quadrant field supplies.

it is safe to shut down one converter and start up the other. Both converters handle power flows in either direction. The field current is independently controlled from a single quadrant controlled rectifier.

Fig. 9.18(b) shows a two-quadrant converter in the armature circuit (E can reverse but not I) with two two-quadrant field converters which allow for field current reversal. The field time constant is much higher than the armature time constant and speed reversal using Fig. 9.18(b) is slower than Fig. 9.18(a). Field forcing can be used by applying a very high voltage to the field circuit to force the current up or down quickly as the case may be and then reducing the applied voltage as the current reaches the desired value.

Field current reversal is cheaper than armature current reversal because the field power levels even in large machines of several MW will only be a few kilowatts. The field time constant may typically be 2 seconds while that of the armature may only be of the order of 0.1 second. Field forcing can improve the field current reversal but it can never better the armature current reversal time.

Unless power-factor on the a.c. side is a problem a static Ward–Leonard system would be used in preference to the original M–G set system mentioned in Chapter 7. Noise may also be a problem with the static system as a result of ripple voltages on the d.c. side. Otherwise the static system is cheaper, more efficient and less bulky than the corresponding M–G set.

Static Ward–Leonard systems are used as elevator drives, mine winding drives, steel rolling mill drives and paper mill drives. The sets range in size from several tens of kW in elevator drives to several MW in mine winding and rolling mill drives. The rolling mill in particular requires rapid reversal and the paper mill requires very close speed control to prevent breaks occurring.

9.8 VARIABLE SPEED A.C. DRIVE SYSTEMS

Both synchronous and induction motors are being used in conjunction with static frequency changers. Although the capital cost is obviously higher than for the machine itself greater versatility is achieved with respect to control of speed, torque, power factor and efficiency.

Fig. 9.19(a) shows a converter fed synchronous motor drive.

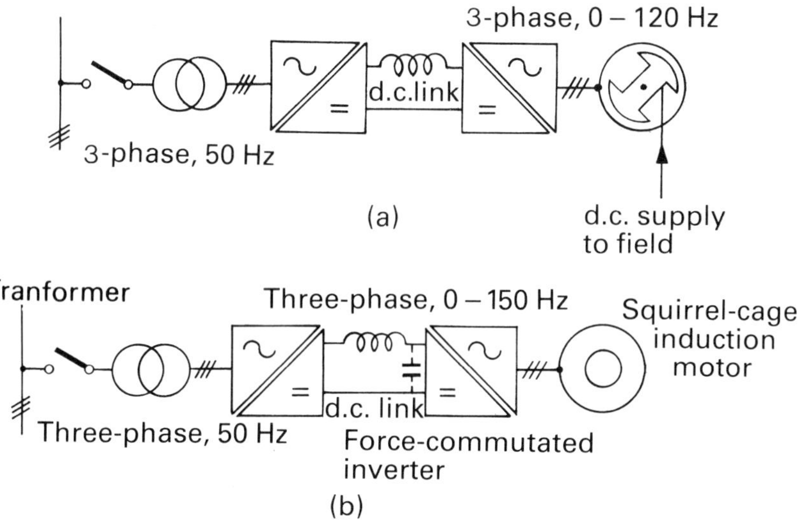

Fig. 9.19(a) A converter fed synchronous motor drive and (b) a squirrel-cage motor fed at variable frequency from a force-commutated inverter.

Both converters are line commutated and can operate as rectifiers or inverters. This allows power flow in both directions although the current direction remains unchanged. (Voltage reverses.) The d.c. link contains a smoothing reactor but no capacitor. This makes the right-hand converter current fed during motor operation rather than voltage fed, i.e. current rather than voltage is constant in the d.c. link. The field current to the rotor is controlled in conjuction with the firing of the converter circuits which allows precise control of the position and amplitude of the B_s and B_r flux density distributions to give the desired torque, speed and power factor. Feedback signals are taken from the rotor position or from the machine stator voltage. Using thyristors there is no upper limit to the power range of this type of system and there are many 10 MVA systems already commissioned. The upper frequency limit is of the order of 120 Hz. When the machine is stationary the machine side converter has no line voltage available for commutation but simple methods have been found to overcome this problem which allow starting against quite high load torques. The main applications for this type of converter system are for pumps, extruders and compressors where precision speed control is required. The ratings are generally in excess of 1 MW. Boiler feed-pumps are increasingly employing this drive making use of the top frequency range to run at 5000 to 6000 rev/min without the need for gearing. Another interesting application is in the starting and running up to speed of M–G sets in pumped storage schemes, synchronous condensers, gas turbine sets and large compressors for blast furnaces and wind tunnels.

Cycloconverter driven synchronous motors are being used for very low speed high power applications in cement mill drives. The mill when running must be set at between 13 and 15 rev/min and much lower speeds are required when the grinding media and armoured plating are being replaced. A multipole synchronous motor driven by a cycloconverter operating at low output frequency can provide a four-quadrant drive which does not require gearing. A typical rating is 10 MW and the near sinusoidal machine current with this type of converter give excellent operation at low speed.

The cycloconverter driven synchronous machine is a serious competitor for the four-quadrant Ward–Leonard d.c. drive in conveying machinery, rolling mills and propeller drives in ice-breakers. It is available in ratings beyond the reach of d.c. machines (greater than 10 MW) and can be used in environments where the maintenance requirements or fire hazards of a commutator render the d.c. machine unsuitable.

The asynchronous or induction motor can be employed in several modes incorporating different converter types. Fig. 9.19(b) shows a squirrel-cage motor fed by a forced commutated inverter. The d.c. link may be current or voltage controlled depending on the presence of the capacitor shown dotted on Fig. 9.19(b). Speed control is effected by varying the inverter frequency keeping the ratio volts/Hz constant for a voltage fed inverter or keeping the stator current constant for a current fed inverter. The latter scheme will reduce the volts/Hz at low frequencies and light loads and this will improve the power factor and efficiency. The constant volt/Hz regime gives poor efficiency and power factor at low speed. This type of drive is available in sizes up to 1 MW using PWM inverters at the higher powers in preference to the cheaper six-step inverters. Considerable interest is being shown in the use of this drive to replace the d.c. series traction drive. Several railway systems are experimenting with inverter controlled squirrel-cage induction motors driving their cars and it seems highly likely that future electric locomotives will be equipped in this way.

In the case of slip-ring induction motors power electronics can be used to absorb or supply power at slip frequency to the rotor. In Chapter 6 we saw that absorbing power in added rotor resistance decreases the rotor speed for a given torque value. Instead of absorbing power in a resistor which is wasting energy as heat we can inject into the rotor coils a voltage of appropriate phase and frequency to act as a power source or sink. This slip power is either taken from or returned to the three-phase mains via converters. Taking power out of the rotor coils will result in a decrease in speed whereas supplying power to the rotor coils will cause an increase in speed.

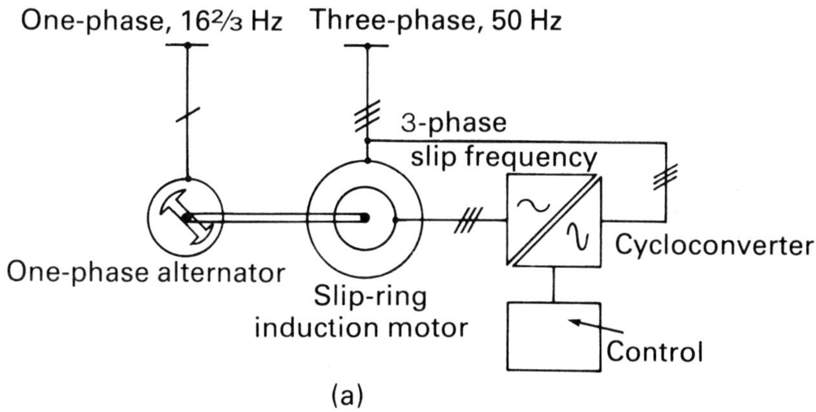

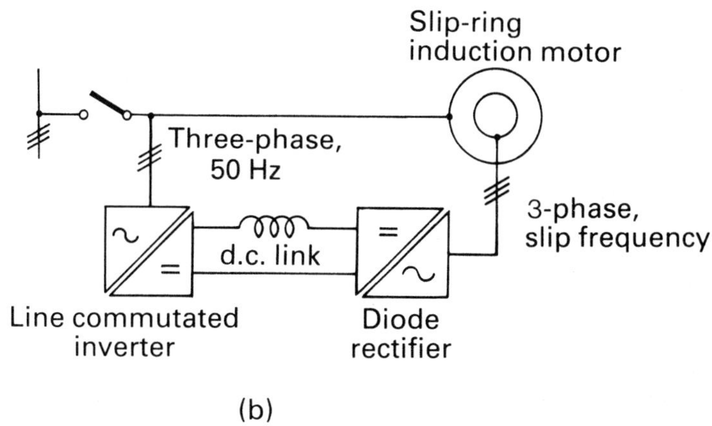

Fig. 9.20(a) A cycloconverter controlling slip power in a slip-ring induction motor and (b) controlling induction motor speed by feeding slip-power back into the mains via a line commutated inverter.

By this technique the motor can be made to run below or above synchronous speed. Provided the speed range is small the slip powers involved are small and the converters can have low ratings. Fig. 9.20(a) shows a particular application of such a scheme. A number of European countries (Austria, Switzerland, Germany) have railway supply systems which operate at 50/3 Hz to supply single-phase universal motors in their electric locomotives. The frequency conversion system consists of a six-pole induction motor driving a two-pole single-phase alternator. In order to generate 50/3 Hz on the railway system the induction motor must rotate at synchronous speed and this is only possible if d.c. is applied to the rotor coils. If, however, the 50 Hz supply frequency deviates then the rotor will require power to be supplied or possibly absorbed at a very low frequency provided the 50 Hz system deviations are small. A cycloconverter is ideal for this purpose and by controlling the phase of the rotor currents with respect to rotor position the power factor on the 50 Hz side can be set to any desired value. Power can be transmitted in either direction as the need arises. Typical unit ratings lie between 8 and 80 MVA with the converter rating being a small proportion of this figure. This type of drive is called a static 'Scherbius' system after the inventor of the original scheme which utilised several machines instead of converters.

Where only sub-synchronous speeds are required a simpler converter can be used with simple diode rectifiers on the rotor side and an inverter on the mains side. Fig. 9.20(b) shows such a system. Provided the speed control range is small then the converter rating can be small. Drives of this type up to 15 MW ratings have been used to drive boiler feed pumps.

Like the d.c. machine the slip-ring induction motor has the deficiencies inherent in rubbing contacts. They are unsuitable for use in say the chemical industry where sparks could be dangerous and in addition the brushes need frequent maintenance.

Reference [9] gives more detailed information on a.c. drive systems.

9.9 CONCLUSIONS

New semiconductor power devices will further increase the controllability and versatility of drive systems. The continuing decrease in semiconductor costs tilts the balance ever more towards the simpler machine and more complicated electronics to give it the required speed/torque characteristic. But there are limits to the process. The only converter which approaches the performance of the d.c. machine commutator is the cycloconverter and it too is a pretty expensive item. No a.c. drive system can as yet match the 30/1 speed range at constant torque which is available from a d.c. drive system. The d.c. machine cannot yet be consigned to the historical dustbin.

Power-electronics has opened up interesting new possibilities in the operation and design of electrical machines. It has also generated a new jargon all of its own which hopefully the reader can now understand. There is little doubt that this application area will expand not only because of the increased versatility it puts at the disposal of the drives engineer but also because of the increased efficiency which in most cases accompanies this versatility.

Appendices

APPENDIX A

Delta-Star Transformation

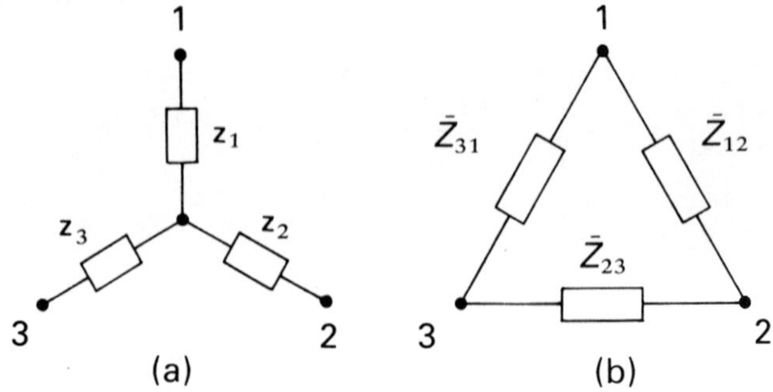

Figure A.1 Star-Delta transform. (a) Star impedances and (b) delta impedances.

The impedance between any pair of terminals is to be the same for the networks shown in Fig. A.1.

$$\bar{z}_1 + \bar{z}_3 = \bar{Z}_{31}(\bar{Z}_{12} + \bar{Z}_{23})/\bar{Z}_{12} + \bar{Z}_{23} + \bar{Z}_{31} \tag{A.1}$$

$$\bar{z}_3 + \bar{z}_2 = \bar{Z}_{23}(\bar{Z}_{31} + \bar{Z}_{12})/\bar{Z}_{12} + \bar{Z}_{23} + \bar{Z}_{31} \tag{A.2}$$

$$\bar{z}_2 + \bar{z}_1 = \bar{Z}_{12}(\bar{Z}_{23} + \bar{Z}_{31})/\bar{Z}_{12} + \bar{Z}_{23} + \bar{Z}_{31} \tag{A.3}$$

To find $\bar{z}_1$ in terms of the delta connected impedance add equations (A.1) and (A.3), subtract equation (A.2) and then divide by 2. This gives

$$\bar{z}_1 = \frac{\bar{Z}_{12}\bar{Z}_{31}}{\sum \bar{Z}} \tag{A.4}$$

where $\sum \bar{Z} = \bar{Z}_{12} + \bar{Z}_{23} + \bar{Z}_{31}$. Similarly

$$\bar{z}_2 = \frac{\bar{Z}_{23}\bar{Z}_{12}}{\sum \bar{Z}} \tag{A.5}$$

and

$$\bar{z}_3 = \frac{\bar{Z}_{31}\bar{Z}_{23}}{\sum \bar{Z}} \tag{A.6}$$

Note that for $\bar{z}_\alpha$ the numerator on the right-hand side of equations (A.4), (A.5) and (A.6) contains the terms which have α in their subscripts, i.e. the delta impedances connected to terminal α. ($\alpha = 1$, 2 or 3.)

Star-Delta transformation

This is more simply evaluated in terms of admittances as follows.

Between 1 and 2 with 2 and 3 short-circuited

$$\bar{Y}_{12} + \bar{Y}_{31} = \bar{y}_1(\bar{y}_2 + \bar{y}_3)/(\bar{y}_1 + \bar{y}_2 + \bar{y}_3) \tag{A.7}$$

Between 2 and 3 with 3 and 1 short-circuited

$$\bar{Y}_{23} + \bar{Y}_{12} = \bar{y}_2(\bar{y}_3 + \bar{y}_1)/(\bar{y}_1 + \bar{y}_2 + \bar{y}_3) \tag{A.8}$$

Between 3 and 1 with 1 and 3 short-circuited

$$\bar{Y}_{31} + \bar{Y}_{23} = \bar{y}_3(\bar{y}_1 + \bar{y}_2)/(\bar{y}_1 + \bar{y}_2 + \bar{y}_3) \tag{A.9}$$

Add equations (A.7) and (A.8), subtract equation (A.9) and then divide by 2 to get

$$\bar{Y}_{12} = \bar{y}_1\bar{y}_2/\sum \bar{y} \tag{A.10}$$

where $\sum \bar{y} = \bar{y}_1 + \bar{y}_2 + \bar{y}_3$. Similarly

$$\bar{Y}_{23} = \bar{y}_2\bar{y}_3/\sum \bar{y} \tag{A.11}$$

and

$$\bar{Y}_{31} = \bar{y}_3\bar{y}_1/\sum \bar{y} \tag{A.12}$$

Note that the subscripts on the delta admittance $\bar{Y}_{\alpha\beta}$ appear on the numerator of the right-hand side as $\bar{y}_\alpha\bar{y}_\beta$.

APPENDIX B

Regulation

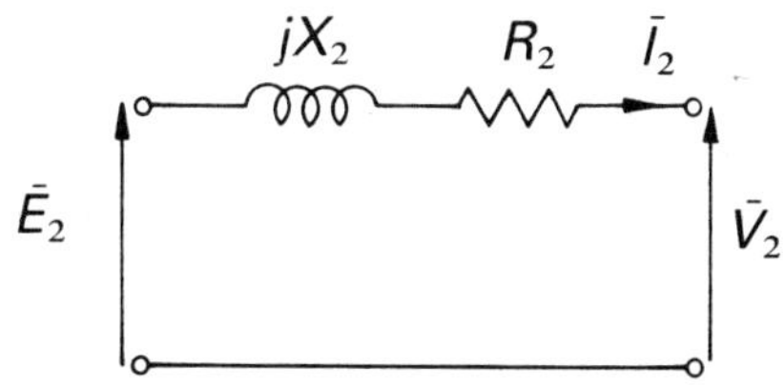

Figure B.1 Circuit diagram for Appendix B.

From Fig. B.1

$$\bar{E}_2 = \bar{V}_2 + \bar{I}_2(R_2 + jX_2) \tag{B.1}$$

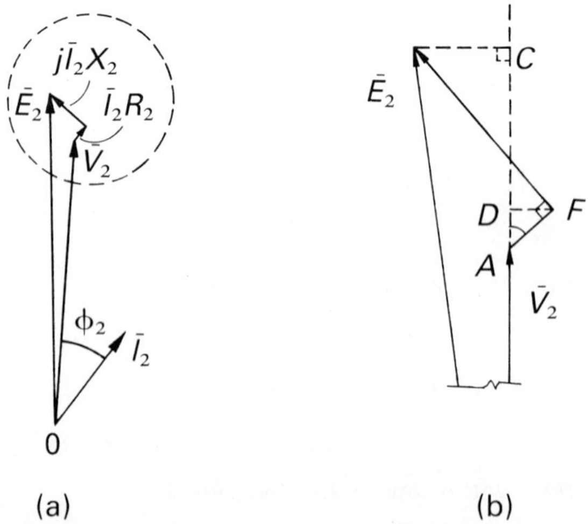

Figure B.2(a) Phasor diagram for the circuit shown in Fig. B1 and (b) enlargement of phasor diagram inside dotted circle of (a).

Fig. B.2(a) shows the phasor diagram for typical values of R_2 and X_2 resulting in a small angular difference between $\bar{E}_2$ and $\bar{V}_2$. The part of the diagram inside the dotted circle is shown enlarged in Fig. B.2(b). In Fig. B.2(b)

$$\begin{aligned} \text{AC} &= \text{AD} + \text{DC} \\ &= \text{AF} \cos \phi_2 + \text{FG} \sin \phi_2 \\ &= I_2R_2 \cos \phi_2 + I_2X_2 \sin \phi_2 \end{aligned} \tag{B.2}$$

Because of the small angular difference between $\bar{E}_2$ and $\overline{OC}$ then

$$E_2 \doteqdot OC$$

and hence regulation

$$\frac{E_2 - V_2}{E_2} = \frac{I_2R_2 \cos \phi_2 + I_2X_2 \sin \phi_2}{E_2} \tag{B.3}$$

Regulation in power transformers is of the order of a few per cent and the denominator of equation (B.3) is quite often taken to be the rated voltage instead of the open-circuit voltage. In Example 3.2 this alternative method of calculating regulation gives an answer of 3.5 per cent in comparison to the exact figure of 3.4 per cent.

Note that equation (B.2) is for a lagging phase angle. If ϕ_2 had been leading then

$$E_2 - V_2 = I_2R_2 \cos \phi_2 - I_2X_2 \sin \phi_2 \tag{B.4}$$

APPENDIX C THE TWO-PHASE INDUCTION MOTOR

Equations (6.27) are

$$\begin{aligned} \bar{V}_a &= \bar{E}_a + \bar{I}_a \bar{Z}_a \\ \bar{V}_b &= \bar{E}_b + \bar{I}_b \bar{Z}_b \end{aligned} \tag{C.1}$$

Rewriting these in matrix form gives:

$$\begin{bmatrix} \bar{V}_a \\ \bar{V}_b \end{bmatrix} - \begin{bmatrix} \bar{E}_a \\ \bar{E}_b \end{bmatrix} = \begin{bmatrix} \bar{Z}_a & 0 \\ 0 & \bar{Z}_b \end{bmatrix} \begin{bmatrix} \bar{I}_a \\ \bar{I}_b \end{bmatrix} \tag{C.2}$$

From equation (6.24) we have also that

$$\begin{bmatrix} \bar{I}_a \\ \bar{I}_b \end{bmatrix} = \begin{bmatrix} 1 & 1 \\ jn & -jn \end{bmatrix} \begin{bmatrix} \bar{I}^+ \\ \bar{I}^- \end{bmatrix} \tag{C.3}$$

and

$$\begin{bmatrix} E_a \\ E_b \end{bmatrix} = \begin{bmatrix} 1 & 1 \\ j/n & -j/n \end{bmatrix} \begin{bmatrix} \bar{E}^+ \\ \bar{E}^- \end{bmatrix} \tag{C.4}$$

We now wish to transform equation (C.2) into quantities related to the forward and backward fields by using equations (C.3) and (C.4). Equation (C.2) can now be rewritten as:

$$\begin{bmatrix} \bar{V}_a \\ \bar{V} \end{bmatrix} - \begin{bmatrix} 1 & 1 \\ j/n & -j/n \end{bmatrix} \begin{bmatrix} \bar{E}^+ \\ \bar{E}^- \end{bmatrix} = \begin{bmatrix} \bar{Z}_a & 0 \\ 0 & \bar{Z}_b \end{bmatrix} \begin{bmatrix} 1 & 1 \\ jn & -jn \end{bmatrix} \begin{bmatrix} \bar{I}^+ \\ \bar{I} \end{bmatrix} \tag{C.5}$$

If we pre-multiply both sides by the inverse of

$$\begin{bmatrix} 1 & 1 \\ j/n & -j/n \end{bmatrix}$$

then the coefficients of

$$\begin{bmatrix} \bar{E}^+ \\ \bar{E}^- \end{bmatrix}$$

will become a unit matrix.

$$\therefore \ \frac{1}{2}\begin{bmatrix} 1 & -jn \\ 1 & jn \end{bmatrix}\begin{bmatrix} \overline{V}_a \\ \overline{V}_b \end{bmatrix} - \begin{bmatrix} \overline{E}^+ \\ \overline{E}^- \end{bmatrix} = \frac{1}{2}\begin{bmatrix} 1 & -jn \\ 1 & jn \end{bmatrix}\begin{bmatrix} Z_a & 0 \\ 0 & Z_b \end{bmatrix}\begin{bmatrix} 1 & 1 \\ jn & -jn \end{bmatrix}\begin{bmatrix} \overline{I}^+ \\ \overline{I}^- \end{bmatrix} \tag{C.6}$$

Let us define $\overline{V}^+$ and $\overline{V}^-$ as

$$\overline{V}^+ = \tfrac{1}{2}(\overline{V}_a - jn\overline{V}_b), \qquad \overline{V}^- = \tfrac{1}{2}(\overline{V}_a + jn\overline{V}_b)$$

Hence from equation (C.6)

$$\begin{bmatrix} \overline{V}^+ \\ \overline{V}^- \end{bmatrix} = \begin{bmatrix} \overline{E}^+ \\ \overline{E}^- \end{bmatrix} + \frac{1}{2}\begin{bmatrix} \overline{Z}_a + n^2\overline{Z}_b & \overline{Z}_a - n^2\overline{Z}_b \\ \overline{Z}_a - n^2\overline{Z}_b & \overline{Z}_a + n^2\overline{Z}_b \end{bmatrix}\begin{bmatrix} \overline{I}^+ \\ \overline{I}^- \end{bmatrix} \tag{C.7}$$

Equations (C.7) are used in non-matrix form as equation (6.28).

References

[1] Hindmarsh, J. 'Electrical machines and their applications', Pergamon Press.

[2] Say, M. G. *Alternating Current Machines*, Pitman Publishing Ltd.

[3] Williams, F. C. and Laithwaite, E. R. 'The action of the single phase induction motor', *Journal of the Institution of Electrical Engineers*, 1961, Vol. 7, Part 9, pp. 550–553.

[4] Cullwick, E. G. *The Fundamentals of Electromagnetism*, Cambridge University Press.

[5] Say, M. G. and Taylor, E. O. *Direct Current Machines,* Pitman, 1980.

[6] Oshima, K. Superconducting magnetic levitation train project in Japan, *IEEE Transactions on Magnetics*, Vol. MAG17, No. 5, September 1981.

[7]

'Switching lots of watts at high speeds, *IEEE Spectrum*, December 1981.

[8]

'Future trends in power semiconductors', *IEE, Electronics & Power*, May 1982.

[9] Bose, B. K. *Adjustable Speed A.C. Drive Systems*, New York, IEEE Press, 1981.

[10] Sen, P. C. *Thyristor d.c. Drives*, John Wiley & Sons, 1981.

[11] Acarnley, P. P. *Stepping Motors*, Peter Peregrinus London, 1982.

Index